Thank you for downloading this compilation of abstracts submitted for the 2015 CMOS Emerging Technologies Research conference.

Other Ebooks from CMOSET Research

Power Line Communication Technologies for Smart Grids, Smart Cars, and Smart Homes

A comprehensive study of Power Line Communications (PLC) technologies and their applications for smart grids, smart cars, and smart homes. This book covers the history and future of PLC and its applications, including the use of PLC for broadband access and in vehicular networks. It also includes chapters on equalization techniques and resource management and allocation techniques for PLC, and touches on the use of PLC for integrating green energy sources into the traditional power infrastructure.

Conference Presentation Slides on Google Books

We are pleased to offer public access to the final programs and presentation slides from each of our previous events. The programs are provided in PDF format, and the presentation slides are published in downloadable ebooks on Google Books.

These ebooks present only a sampling of the topics discussed at the conference. Many speakers present proprietary or not-yet-published material that cannot be distributed in this format, and of course the in-person conversations range far beyond the material contained in these slides. The best way to experience the opportunities for collaboration and cooperation that come from our conferences is to join us.

CMOSET Research Symposia

CMOS Emerging Technologies was founded in 2006 to provide researchers and industry representatives in the high-tech sector with an opportunity to discuss new and exciting developments in all areas of high technology. Our annual conferences provide companies and academic institutions with an international stage for showcasing their technology, innovations, products and services. Our participants come from around the world and represent every segment of high-tech, from VLSI to green energy to wireless communications to photonics. Together, we create a stimulating common ground for exploring collaborations and encouraging discussions on emerging technologies.

The CMOS Emerging Technologies Research conferences are held annually, and we sincerely hope to see you at the next one.

CMOS Emerging Technologies Research
www.cmosetr.com

Abstracts

Track P: Plenary Sessions

Session P1: Plenary I

Jan Rabaey, University of California, Berkeley (jan@eecs.berkeley.edu)

The Return of Neuro-Inspired Computing - Why Now?

Barring technologies surprises, alternative design strategies may be necessary if continued scaling of functionality in terms of size and energy is to be obtained. Neuro-inspired computing is one possible direction to be considered. The brain is an amazingly complex and efficient machine. While it may not be considered "general purpose" in terms of its computational capabilities, it performs a set of functions such as feature extraction, classification, synthesis, recognition, learning, and higher-order decision-making amazingly well.

Today, it is realized that neuro-inspired computing may be a perfect match to the properties of the emerging nano-scale devices: it thrives on randomness and variability, processing is performed in the continuous or discrete domains, and massive parallelism, major redundancy and adaptivity are of essence. Computational paradigms inspired by neural information processing hence may lead to energy-efficient, low-cost, dense and/or reliable implementations of the functions the brain excels at. In this presentation, we will explore various means on how the interaction between neuroscience and information technology may lead to an exciting future.

Giovanni De Micheli, École Polytechnique Fédérale de Lausanne (giovanni.demicheli@epfl.ch)

3-Dimensional Devices: Models and Design Tools

Three-dimensional devices, from FinFETs to NanoWire FETs, are enabling us to downscale transistor dimensions to 10 nanometers and below, by providing us with better electrostatic control. Moreover, such devices can be enhanced by additional structures, such as multiple gates, that provide extended functionality, including polarity control, thus increasing the computational density on silicon.

Enhanced device features must be matched by corresponding design tools and methods that exploit these characteristics to realize efficiently computing functions. In particular, such enhanced devices can realize computational fabrics for data paths that are superior to standard CMOS.

Thomas Theis, Semiconductor Research Corporation (Thomas.Theis@src.org)

Emerging Devices for Logic: Can a Low-Voltage Switch be Fast?

Since roughly 2003, the decreasing ability to reduce supply voltages, combined with constraints on power, has forced designers to limit clock frequencies even as devices have continued to shrink. The device physics must change in a fundamental way if we are to realize faster digital logic with very low power dissipation. Can any of today's "emerging" devices compliment and then someday replace CMOS transistors? III-V tunneling FETs (TFETs) promise to open a new ultra-low-power design space. Nanomagnetic devices may allow memory and logic to be combined in novel ways. And newer, more promising device concepts continue to emerge.

Richie Przybyla, Chirp Microsystems (rjp@eecs.berkeley.edu)

Ultrasonic 3D Rangefinder on a Chip

As the number of battery powered internet devices explodes, there is an increasing need to build contextual awareness into these devices so they have some idea of what is happening around them. A key feature of a sensor used for contextual awareness is that it should be kept always on so that it can wake the host microprocessor when a target event occurs. At Chirp Microsystems, we are bringing time-of-flight MEMS ultrasound to market to enable microwatt-level contextual awareness which can be used in applications ranging from proximity sensing to 3D gesture recognition. I will present measured results from an ultrasonic 3D rangefinder system which uses an array of AlN MEMS transducers and custom readout electronics to localize targets over a +/-45 degree field of view up to 1m away. The 0.18 µm CMOS readout ASIC comprises 10 independent channels with separate high voltage transmitters, readout amplifiers, and ADCs. Power dissipation is 400 µW at 30 fps, and scales to 10 µW/ch at 10 fps.

David Cumming, University of Glasgow (David.Cumming.2@glasgow.ac.uk)

Integrated Circuits for Sensors

CMOS is at the heart of modern computing and communications technology. However, it is possible to think of the technology for making CMOS microelectronics as a material system with far more diverse possibilities. In this talk I will present results on the hybridisation of CMOS and nanotechnology for a range of sensor applications. These will include ion sensing for chemical and biological imaging and the all electronic next generation sequencing system. I will then explore the possibilities for hyperspectral and terahertz imaging based on the integration plasmonics and metatmaterial structures on CMOS.

COFFEE BREAK (GEORGIA FOYER)

****

Session P2: Plenary II

Eli Yablonovitch, University of California, Berkeley (eliy@eecs.berkeley.edu)

Search for Lower Voltage Switch with Nano-electronic, Nano-mechanical, Nano-photonic, and Nano-magnetic Approaches

Kevin Fu, University of Michigan (kevinfu@umich.edu)

Medical Device Cybersecurity

David Ricketts, North Carolina State University (dricket@ncsu.edu)

State of the Art and Future of High-Speed Wireless Transceivers: Opportunities and Challenges for Near-THz Communication

Esther Rodriguez Villegas, Imperial College London (e.rodriguez@imperial.ac.uk)

Interfacing Biology and Circuits

Wearable and implantable electronic systems for monitoring physiological signals are heavily constrained both by weight and volume specifications. However these constraints can considerably differ depending on the intended specific application of those systems, even if the nature of the signal that is being measured is the same. Because of this, the optimum design approach is not always a generic one but rather one that takes into account the application related requirements. This talk will review different applications of wireless brain monitoring systems and will provide circuit design recommendations to improve the power consumption and overall form factors for these systems.

Daniel Hammerstrom, DARPA (daniel.hammerstrom@darpa.mil)

Unconventional Processing of Signals for Intelligent Data Exploitation

Today's Defense missions rely on massive amounts of sensor data collected by intelligence, surveillance and reconnaissance (ISR) platforms. Not only has the volume of sensor data increased exponentially, there has also been a dramatic increase in the complexity of analyses required for applications such as target identification and tracking. The digital processors used for ISR data analysis are limited by on-platform power constraints, potentially limiting the speed and type of data analysis that can be done. Furthermore, as Moore's Law slows down, power scaling has more or less stopped.

The UPSIDE program seeks to break the status quo of digital processing with methods of video and imagery analysis based on the physics of MS CMOS and eventually nanoscale devices. UPSIDE processing is non-digital and fundamentally different from current digital processors and the power and speed limitations associated with them. Unlike traditional digital processors that operate by executing specific instructions to compute, UPSIDE computational arrays will rely on a higher level computational element based on probabilistic inference.

COFFEE BREAK (GEORGIA FOYER)

****

Session P3: Plenary III

John Rogers, University of Illinois at Urbana-Champaign (jrogers@illinois.edu)

Silicon Integrated Circuits Constructed with Materials that are Completely Soluble in Water and Biofluids

Anthony Guiseppi-Elie, Clemson University (aguisep@clemson.edu)

with O. Karunwi and F. Alam

Functionalized SWCNT as Chemoresistors for the Electronic NOSE™

There is need for advanced synthetic molecular chemoreceptors in the design and development of all electronic Natural Olfactory Sensor Emulators® (NOSEs). Carbon nanotubes with their semiconducting to metallic properties and relative ease of non-covalent and covalent functionalization are well suited for chemoreceptor use in e-Noses. However, the choice of metal-to-SWCNT contact for the most efficient chemoresistive response remains dubious. A functionalized SWCNT rationalized on the basis of covalent chemical modification to impart an electron-withdrawing functional group through photobromination has resulted in brominated-SWCNT (SWCNT-Br) that may serve as a chemoreceptor and facilitate the development of highly selective and sensitive chemical and biological sensor arrays by an approach that mimics the mammalian olfactory system – "electronic nose". SWCNT-Br was dispersed in m-cresol and precursor pristine SWCNT (SWCNT-p) in DMF and both cast onto the interdigit space of microlithographically fabricated interdigitated microsensor electrodes (IME) of platinum or gold. The IME 1025-M-Pt or Au devices were 10 μm line and space with 25 fingers of M = Pt or Au (100nm)/TiW(20nm) patterned on borosilicate glass. Before casting, IME devices were solvent cleaned and chemically modified with octadecyltrichlorosilane (OTS), cathodically cleaned (20 cycles, 100 mV/s, PBS 7.2) to remove adsorbed silane from the digits, rinsed isopropyl alcohol and dried in flowing argon. The resulting devices (receptors) were characterized by two–electrode impedance spectroscopy, I-V characterization, temperature dependent impedimetry, and multiple scan rate cyclic voltammetry (MSRCV) in PBS supported $[Fe(CN)6]3-/4-$. The chemo-impedimetric responses of these devices to four vapors; ethanol, methyl isoamyl ketone (MIAK), toluene and octane (representing alcohols, ketones, aromatics and aliphatics) were also studied. SWCNT-Br were more conductive than SWCNT-p. Gold devices were found to support larger, more reproducible and more repeatable chemoresistive responses.

Vladimir Stojanovic, University of California, Berkeley (vlada@berkeley.edu)

Silicon Photonics for VLSI Systems

In this talk we will present the latest results on the integration of silicon-photonic interconnects into a monolithic (45nm SOI logic process and bulk CMOS memory periphery process) and a 3D platform. We also illustrate some critical aspects of this technology that need to be addressed from integration, circuits and systems side. These breakthroughs pave the way for orders of magnitude improvement in performance of photonically-enhanced VLSI systems. Moreover, just like integrating the inductor into CMOS in 1990s revolutionized the RF design and enabled mobile revolution, integration of silicon-photonic active and passive devices with CMOS is greatly positioned to revolutionize a number of analog and mixed-signal applications – low-phase noise signal sources and large bandwidth, high-resolution ADCs, to name a few.

Ajit Khosla, Concordia University (khosla@gmail.com)

Nanocomposite Polymers for Micro-Nano-Systems and Health Care

Nanotechnology is set to have a major impact on our society and the way we live. It has already brought advances such as self-cleaning windows, high efficiency solar cells, ice accumulation resistant aircraft exteriors, coatings for pipelines, and nanoparticle reinforced nanocomposites that are lighter and tougher than steel. Medical nanotechnological breakthroughs in areas such as drug delivery and cancer detection are also on the horizon, and have the potential to revolutionize our health care system.

It is clear that adhesion between polymers and metals needs molecular bonding. Since molecular forces are relatively short range, close contact and good wetting conditions are required.

This talk focuses on a particular type of nanotechnological breakthrough: nano-particle doped micropatternable multi-functional polymers. Polymers are inherently electrically insulating and non-magnetic but these properties can be modified by the introduction of conducting and/or magnetic nanoparticles in the polymer matrix. This enables the polymers to retain their inherent benefits (ease of fabrication, cost, mechanical and surface properties while being rendered functional in some way, e.g., being conductive, magnetic, or mechanically active. Adhesion between polymers and metals needs molecular bonding. Since molecular forces are relatively short range, close contact and good wetting conditions are required. Common practice that improves this situation will also be discussed.

This talk will cover the fabrication and process technology of micropatternable multifunctional nanocomposite polymers/resists for M- (micro-) and N- (nano) EMS (electromechanical systems), and discuss their employment in a number of applications, such as: shape-conformable micro-electrodes for applications in tissue tomography, wearable sensors, conductive threads, flexible 3-D printed electronics and solar cells, will also be discussed.

COFFEE BREAK (GEORGIA FOYER)

****

Track C: Circuit Advances & Emerging Technologies

Session C1: Digital Circuits and Systems

Bob Merritt, Convergent Semiconductors (bobm@convergentsemiconductors.com)

Does the Digital Revolution Lead to a Replay of the Industrial Revolution?

We debate the rate at which new technologies will continue to become available, but we don't doubt that we will continue to see rapid technological advances.

Is it also time to consider the future social implications as the pace of technology development continues?

There are two scenarios challenging our contemporary view of society that we can use to measure our technical advances relative to a disruptive social impact. One scenario is the merging of man and machine into a single entity. The other scenario is the possibility of physical communications channels directly between human brains.

This session presents the current status of those two scenarios, reviews the social impact of the Industrial Revolution, and speculates on the likelihood of a similar social restructuring.

Scott Smith, North Dakota State University (scott.smith.1@ndsu.edu)

The Future of Asynchronous Logic

ITRS 2012 states that asynchronous circuits account for 22% of logic within the multi-billion dollar semiconductor industry, and predicts that this percentage will more than double over the next 10 years. Asynchronous logic has been around for the past 50+ years; but, until recently, synchronous circuits have been good enough to meet industry needs. However, as transistor size continues decreasing, asynchronous circuits are being looked to by industry to solve power dissipation and process variability issues. This talk will discuss the state-of-the-art of asynchronous logic, how asynchronous circuits are currently being utilized in industry, and the future of asynchronous logic.

Zhengya Zhang, University of Michigan (zhengya@umich.edu)

Algorithm-Circuits Co-Design to Enable Low-Power and High-Speed Channel Decoders

Forward error correction has been used to save transmit power for a reliable delivery, but the decode power of the best capacity-approaching codes can be prohibitive due to circuit complexities. The rising decode power poses obstacles to future applications. Addressing this problem requires algorithm-circuits co-design. We show that significant memory and logic power can be saved by an efficient embedded memory and a runtime gating strategy to exploit algorithm characteristics. The results are demonstrated in chip designs for state-of-the-art channel decoders.

Farid Najm, University of Toronto (f.najm@utoronto.ca)

Verification and Design of the Power Delivery Network in VLSI Circuits

The power distribution network of an IC must be checked during the design process to ensure that supply voltage fluctuations meet the design specs. One way of doing this is by simulation, but it requires knowledge of the circuit currents, which are hard to specify. In many cases, they may be simply unknown because the circuit itself may not yet be specified. We will review methods for vectorless verification, developed over the last decade, for checking the grid in the absence of knowledge of the circuit currents. We also describe recent work on generating specs for grid safety.

Brad Quinton, Invionics Inc. (bradq@invionics.com)

The Changing Landscape of Electronics Design Automation (EDA): Trends, Challenges and Opportunities

The Electronics Design Automation (EDA) industry is changing dramatically. After years of without VC funding, the entire concept of EDA startups is in question. "All-you-can-eat" pricing models from the big three EDA players are encouraging single vendor homogeneity in the design flow. The cost of IC design continues to increase rapidly increasing the size of the "gamble" in each new process node. And, finally, there is a renewed trend towards vertical integration of silicon design at traditional system developers like Apple and Microsoft. These trends are creating challenges in EDA and are widening the gap between new EDA ideas and the commercialization of those ideas. However, there are signs of opportunity, as well, as new platform approaches are emerging that promise to bring the "democratization of development" effect that so dramatically changed the mobile app space to the EDA world.

Nasser Kurd, Intel (nasser.a.kurd@intel.com)

Scalable High Performance, Low Power 22nm Processor

This talk describes the 4th Generation Intel® Core™ processor family (code named Haswell) implemented on Intel® 22nm technology that supports scalable form factors from desktops to fan-less Ultrabooks™. Haswell enabled 50% or more improvements in battery life. Among the technologies that enabled power reduction, sleeker form factor, and improved active power-performance are: new deeper power states with fast entry/exist times, independent voltage/frequency domains with individually controlled voltage frequency points, fully integrating the voltage regulators, optimizing MCP I/O system (1-1.22pJ/b), and improving DDR I/O circuits (40% active and 100X idle power saving).

Session C2: Analog Circuits

Ken Coffman, Fairchild Semiconductor (Ken.Coffman@FairchildSemi.com)

Ten Methods for the Creative Destruction of a Power MOSFET

Semiconductors are ubiquitous in the fabric of our modern technological life; most of us are surrounded by many billions of transistors. Mostly, they quietly and efficiently do their jobs and we take them for granted-barely noticing their existence. However, transistors fail. From the perspective of field engineering support, it often seems as if design engineers intentionally devise methods of their mass destruction. The purpose of this paper is to list ten basic methods of destroying a MOSFET-giving the engineering agents of creative destruction a palette of techniques that can be mixed and matched for maximum exothermic entertainment.

As counterpoint, the engineer who values MOSFETs and wants to create robust designs could, as a start, avoid these ten hazards.

Carlos Galup Montoro, Universidade Federal de Santa Catarina (carlosgalup@gmail.com)
with M. C. Schneider

Ultra-low Voltage CMOS Circuits

The main solution to reduce the energy consumption of electronic circuits is to lower the supply voltage. Theoretically, the minimum supply voltage for a CMOS inverter is $2(\ln2)(kT/q) = 36$ mV at room temperature, as shown by Swanson and Meindl in 1972.

In this paper we analyze a CMOS Schmitt Trigger circuit and show that it can operate below the Meindl low-voltage limit. In the following, we will show that analog circuits such as rectifiers and oscillators can operate with supply voltages below (kT/q). Finally, some ultra-low-voltage circuits for energy harvesting will be presented.

Bal Sandhu, ARM (bal.sandhu@arm.com)

An Ultra-low Power Voltage Regulator for Wireless Sensor Applications

The evolving IoT market, especially sensor nodes for pacemakers and other implantable electronic medical devices, places extreme demands on energy consumption. For wireless sensor nodes, we need to operate the processing blocks at near threshold voltages to minimize the current drawn from the battery or energy harvester. This motivates the design and implementation of an ultra-low power voltage regulator for wireless sensor applications.

Woogeun Rhee, Tsinghua University (wrhee@mail.tsinghua.edu.cn)

Ultra-Wideband Technology for Short-Range Communications

Short-range communication is a fast growing area in modern wireless systems. Ultra-wideband (UWB) communication has received great attention in early 2000 and lost the momentum due to practical issues such as the receiver complexity and interference problems. In this talk, system perspectives and design aspects of the UWB transceiver system are discussed, focusing on two possible areas; one for high data rate applications including Gb/s cm-range pointing communication systems, and the other for ultra-low power applications including high-quality hearing aid devices.

Nicolas Rouger, Centre National de la Recherche Scientifique (nicolas.rouger@g2elab.grenoble-inp.fr)

CMOS Gate Drivers for High Speed - High Voltage Wide Bandgap Power Switches

In the view of driving novel wide bandgap power semiconductor devices such SiC MOSFET or GaN HEMT high voltage transistors, several auxiliary functions are required (floating supply, power gate charge control, isolation, sensors and protections). Such functions are integrated within the gate driver, a dedicated CMOS circuit. Key technologies and design constraints on such gate driver analog circuits will be presented, in the context of high speed driving, High Voltage isolation and high temperature operation. The integrated insulation unit for the transfer of gate signals is highlighted with several designs and characterizations (CMOS coreless transformers or CMOS optical detectors).

COFFEE BREAK (GEORGIA FOYER)

****

Marcel Kossel, IBM (mko@zurich.ibm.com)

Tomlinson-Harashima Precoding (THP) for High-Speed IO Links

Tomlinson–Harashima (TH) precoding is a transmitter equalization technique in which the post-cursor intersymbol interference (ISI) is canceled by means of an infinite impulse response (IIR) filter with modulo (MOD)-based amplitude limitation. TH equalizers are suited for asymmetric links where the transmitter contains the equalization complexity and the receiver is kept simple. We present in this talk several methods to reduce the data rate limiting feedback delay such as pipelining, sub-rate operation and modulo speculation and also show some measurement results from a 8-tap 6b TH Tx in 22nm SOI CMOS technology operated at 10Gb/s in 4-PAM mode.

SeongHwan Cho, KAIST (chosta@ee.kaist.ac.kr)

High-Performance Time-to-Digital Converter Using Time-domain Arithmetic Circuit

In this talk, basic building blocks of time-domain analog arithmetic circuits such as time adder, time amplifier, and time register are introduced. By employing these circuits, energy-efficient time-to-digital converters (TDCs) are implemented such as two-step and pipelined TDCs. The two-step TDC employs a coarse and a fine delay-line based TDC with a pulse-train time amplifier for residue amplification. The pipelined TDC mimics the architecture of the ADC counterpart, and uses time register to store time information. Prototype ICs show excellent performance in resolution, power consumption and speed, leading to state-of-the-art energy-efficiency.

Ramón González Carvajal, Universidad de Sevilla (carvajal@us.es)

High-Performace Op Amp-less Circuits Based on Flipped Voltage Follower

The steady downscaling of MOS transistors improved dramatically the performance of digital integrated circuits, however new challenges in analog design have arisen. Among these challenges, one of the most important is the difficulty of designing high-gain op-amps.

This paper proposes the use of the Flipped Voltage Follower (FVF) cell to avoid the need for classical high-gain op-amps. Taking advantage of the local feedback of the FVF, different Continuous-Time sigma-delta modulators and Low-dropout regulators (LDO) are presented. Experimental results show sigma-delta modulators with extremely small active area and good power efficiency and internally compensated LDO with a good voltage stability.

Fei Yuan, Ryerson University (fyuan@ryerson.ca)

Adaptive Decision Feedback Equalizers for Gbps Data Links

Vamsy Chodavarapu, McGill University (vamsy.chodavarapu@mcgill.ca)

Wide-temperature Range CMOS Interface Circuits for MEMS Sensors

We present a variety of CMOS interface circuits for capacitive MEMS sensors that can function over a wide temperature range between -55°C and 225°C. The circuits are implemented using IBM 0.13μm CMOS technology with 2.5V power supply. Constant-gm biasing technique and other design considerations are applied to mitigate performance degradation at high temperatures. The circuits offer the flexibility to interface with MEMS capacitive sensors with a wide range of the steady-state capacitance values and are designed to provide voltage, frequency and digital outputs. Simulation and experimental results show that these circuits offer high accuracy and stability over the wide temperature range.

Malgorzata Chrzanowska-Jeske, Portland State University (jeske@ece.pdx.edu)

Microelectronics Applications for Harsh Environment

Rohit Mittal, Intel (Rohit.mittal@intel.com)

Session C3: Flexible Electronics

Hanqing Jiang, Arizona State University (Hanqing.Jiang@asu.edu)

Origami Electronics

Origami, creating three-dimensional (3D) structures from two-dimensional (2D) sheets through a process of folding along creases, has been transformed by mathematicians, scientists, and engineers to utilize the folded objects' deformability and compactness in applications. This presentation demonstrates the fabrication of origami electronics that has superb flexible, stretchability and foldability. The fabrication processes here represent an example to utilize mainstream high-temperature processes to fabricate high-performance stretchable electronics. Two examples, namely origami solar cells and origami lithium ion batteries will be presented. It is expected that this work paves the way to explore new and exciting engineering applications of origami.

Chris Bower, X-Celeprint (cbower@x-celeprint.com)

Micro-transfer Printing

Jonathan Rivnay, Centre Microélectronique de Provence (rivnay@emse.fr)

Bio-Related Flexible Organic Electronics

Takeo Someya, University of Tokyo (someya@ee.t.u-tokyo.ac.jp)
with T. Yokota, S. Lee, M. Kaltenbrunner and T. Sekitani

Bionic Skin Using Ultraflexible Organic Electronics

Mechanically flexible devices are expected to open new possibilities in fields of biomedical applications as well as wearable electronics. Especially, conformability, ruggedness, lightweight, biocompatibility, and large-area are all important to create new electronic applications that can be directly mounted on the surface of human skins. Such a bionic skin could be used to monitor medical conditions or to provide more sensitive and lifelike prosthetics. From this viewpoint, ultraflexible organic thin-film devices have attracted much attention recently. Here we report our recent progress of ultraflexible organic thin-film devices, their emerging applications, remaining issues and future prospects.

Fabio Cicoira, École Polytechnique de Montréal (fabio.cicoira@polymtl.ca)
with S. Zhang, P. Kumar, A.S. Nouas, L. Fontaine and H. Tang

Processing of PEDOT:PSS Films for Organic Electrochemical Transistors

Organic electrochemical transistors (OECTs) based on the conducting polymer poly(3,4-ethylenedioxythiophene) doped with poly(styrenesulfonate) (PEDOT:PSS) are currently employed for several bioelectronic applications in vitro and in vivo. However, little is known about the changes PEDOT:PSS films undergo during device operation, due to interactions with the interfacing liquids. In this letter, we investigate the changes induced by immersion of PEDOT:PSS films, processed by spin coating from different mixtures, in liquids of different polarity. We found that the film thickness decreases upon immersion in polar solvents, while the electrical conductivity remains unchanged. The decrease in film thickness is minimized via the addition of a cross-linking agent to the mixture used for the spin coating of the films.

Ana Claudia Arias, University of California, Berkeley (acarias@eecs.berkeley.edu)

with Y. Khan, C. Lochner, A. Pierre and F. Pavinatto

Large Area Flexible Organic Optoelectronic Sensor

Organic semiconductors developed for lights emitting diodes (LEDs) and photodiodes (OPDs) have been primarily applied to display and solar technologies, due to their potential for large-area roll-to-roll manufacturing at high volumes. These materials are solution processed, compatible with printing technologies and the use of flexible substrates. These attributes also make organic optoelectronics very attractive for medical sensors, where flexibility combined with large areas can result in an improvement of the overall sensor performance. In the past 10 years a lot of resources were used to improve the stability of organic semiconductors in order to meet the lifetime requirements of displays and photovoltaic devices. When compared to the above markets disposable medical sensors have less stringent lifetime requirements on the materials, since these devices would be used only for a few days as opposed to years. Printing technologies also allow the sensor design to match the shape and size of sensor to the user. In this talk, we will discuss the use of organic semiconductor materials and metal nanoparticles in large area medical devices. We demonstrate that vital signs such as pulse, oxygen saturation, temperature and ECG can be accurately monitored using novel printed devices.

Session C4: Microfluidics

Philip Brisk, University of California, Riverside (philip@cs.ucr.edu)

Programmable, Integrated Microfluidic Technology: Automating and Miniaturizing Chemistry and Biochemistry

This talk will introduce a domain-specific programming language, compiler, and runtime environment that enables software to control a "cyber-physical" laboratory-on-a-chip (LoC) based on the principle of electrowetting-on-dielectric. In the most recent generation of electrowetting-based LoCs, integrated sensors and video monitoring equipment form a closed feedback loop with the computer that controls the device. The programming language, compiler, and runtime environment have been designed to enable fast decision-making in real-time in response to sensory feedback. This will allow the users of electrowetting devices, for the first time, to specify long-running complex biological protocols that adjust their behavior in response to sensor feedback provided by the device.

Richard Oleschuk, Queen's University (Richard.Oleschuk@chem.queensu.ca)

with L. Mats, N. Mei, A. Bramwell and R. Young

Droplet Based Microfluidics, and (Super) Hydrophobic Coatings

Droplet based microfluidic platforms are an attractive method to move and manipulate discrete samples and to minimize sample carry over. Droplet movement is accomplished by generating electrostatic forces on an array of electrodes coated with an insulating dielectric, electrowetting on dielectric or EWOD or alternatively through magnetic based actuation. Devices for droplet-based systems rely on a hydrophobic layer to minimize droplet surface wetting (typically a perfluoropolymer) which has presented robustness challenges and limited hydrophobicity for magnetic actuation. Superhydrophobic surfaces derive their properties from a combination of micro and nanostructured surfaces that have been coated with a hydrophobic material. We have utilized fluorinated silica nanoparticles to produce superhydrophobic surfaces to examine both the electrowetting properties of the different silica nano particle based surfaces as well as magnetic droplet actuation with paramagnetic particles.

James Friend, University of California, San Diego (jfriend@eng.ucsd.edu)

Acoustic Microfluidics: from Chip in a Lab to Lab on a Chip

Surface acoustic waves have found new life in microfluidics with an enormously powerful ability to manipulate fluids and suspended particles in open and closed fluid systems. After noting the extreme accelerations possible from acoustic irradiation of a fluid, and noting the breadth of research work in the discipline over the past ten years, more recent developments in thin film and droplet formation and manipulation will be presented, as will novel results in nanoparticle separation, deagglomeration, and orientation. Along the way, the fascinating underlying physics tying together the acoustics, fluid dynamics, and free fluid interface in these systems will be described.

Xinyu Liu, McGill University (xinyu.liu@mcgill.ca)

Paper-Based Microfluidic Nano-Biosensors

This talk will present our recent development of microfluidic paper-based microfluidic biosensors, featuring highly-sensitive working electrodes decorated with semiconductor zinc oxide nanowires, for electrochemical detection of metabolite (glucose) and protein (antigen/antibody) markers. Besides common features of paper-based analytical devices such as low cost, high portability/disposability, and ease of operation, the reported devices have several additional advantages such as higher accuracy/sensitivity, improved device stability, and simple label-free protein detection. Our ongoing efforts on the use of these biosensors for clinical diagnoses will also be discussed.

Craig Priest, University of South Australia (Craig.Priest@unisa.edu.au)

Fluid Behaviour On and In Microstructured Surfaces

The design of microstructured surfaces is vital in the design of microdevices for fluid handling. Here, the phenomenon of "wicking", where a liquid spontaneously fills a surface structure, will be presented in the context of precise thin film formation and chemical detection by spectroscopy. Wicking theory and its present limitations will be outlined for spontaneous filling of micro-pillar arrays by aqueous samples and applications in microscopic fluid control will be given.

COFFEE BREAK (GEORGIA FOYER)

****

James Li, University of Texas (xli4@utep.edu)

with M. Dou

Paper/PDMS Hybrid Microfluidic Biochip for Instrument-free Infectious Diseases Diagnosis

Infectious pathogens often cause serious economic loss and public health concerns throughout the world. One important characteristic of infectious diseases is that they often occur in high-poverty regions, where people cannot afford expensive and bulky equipment. Although numerous polydimethylsiloxane (PDMS) microfluidic devices have been developed for bioapplications, PDMS/paper hybrid systems that take advantage of both substrates are rarely reported. Each device substrate has its own advantages and disadvantages. Herein, we have developed different low-cost PDMS/paper hybrid microfluidic systems that take advantage of both PDMS and paper substrates for rapid and sensitive infectious disease diagnosis, especially in low-resource settings.

Sean Collignon, Massachusetts Institute of Technology (s3412939@student.rmit.edu.au)

Planar Microfluidic Drop Splitting and Merging

Open droplet microfluidic platforms offer attractive alternatives to closed microchannel devices, including lower fabrication cost and complexity, significantly smaller sample and reagent volumes, reduced surface contact and adsorption, as well as drop scalability, reconfigurability, and individual addressability. For these platforms to be effective, however, they require efficient schemes for planar drop transport and manipulation. While there are many methods that have been reported for drop transport, it is far more difficult to carry out other drop operations such as dispensing, merging and splitting.

In this work, we introduce a novel alternative to merge and, more crucially, split drops using laterally-offset modulated surface acoustic waves (SAWs). The energy delivery into the drop is divided into two components: a small modulation amplitude excitation to initiate weak rotational flow within the drop followed by a short burst in energy to induce it to stretch. Upon removal of the SAW energy, capillary forces at the center of the elongated drop cause the liquid in this capillary bridge region to drain towards both ends of the drop, resulting in its collapse and therefore the splitting of the drop. This however occurs only below a critical Ohnesorge number, which is a balance between the viscous forces that retard the drainage and the sufficiently large capillary forces that cause the liquid bridge to pinch.

We show the possibility of reliably splitting drops into two equal sized droplets with an average deviation in their volumes of only around 4% and no greater than 10%, which is comparable to the less than 7% splitting efficiencies associated with electrowetting drop splitting techniques. In addition, we also show that it is possible to split the drop asymmetrically to controllably and reliably produce droplets of different volumes. Such potential as well as the flexibility in tuning the device to operate on drops of different sizes without requiring electrode reconfiguration, i.e., the use of different devices, as is required in electrowetting, therefore makes the present method an attractive alternative to electrowetting schemes.

Session C5: Semiconductor Memories

Dan Lin, Micron (flin@micron.com)

Energy Efficient Memory Interfaces and Magic Behind Hybrid Memory Cube (HMC)

Memory bandwidth is a limiting factor for high-performance computing (HPC). Current double-data rate (DDR) DRAMs follow an evolutionary path and face big challenges to address bandwidth, power and scalability issues. To enable the next 10X leap in memory bandwidth, a three-dimensional memory architecture, Hybrid Memory Cube (HMC), has been demonstrated to change the landscape. Equipped with through-silicon vias (TSVs), the heterogeneous integration between a memory stack and a logic layer greatly increases number of I/Os while simultaneously reduces the distance signals travel. This talk will compare energy-efficient memory interfaces used in various systems, i.e. LPDDR4 and high-BW memory (HBM), and then explore the magic behind HMC systems with sub 1 pJ/b energy efficiency as well as more than 1 Tb/s bandwidth.

Gregory Di Pendina, CEA (Gregory.DiPendina@cea.fr)

NV Logic and Memory Based on Spin Orbit Torque MTJ

Recent research on MRAM-based non-volatile logic and memories mainly considers Spin Transfer Torque Magnetic Tunnel Junctions, which suffers from limitations due to the use of the same writing and reading paths: high read disturb failure rates and endurance aspects. A much more recent technology enables to avoid such issues. Indeed, Spin Orbit Torque Magnetic Tunnel Junctions have 3 terminals, enabling to separate the writing path from the reading path, leading to a quasi-infinite endurance and high reliability. This talk will addresses compact modelling, hybrid process design-kit and flow, novel architectures of SOT-MTJ based standard cells and novel MRAM architecture.

Farnood Merrikh-Bayat, University of California, Santa Barbara (F_merrikhbayat@umail.ucsb.edu)
with M. Prezioso, B. Hoskins, X. Guo, G. Adam and D. Strukov

Artificial Neural Network Implementation based on Emerging Memory Technologies

Building artificial neural networks (ANNs) capable of emulating the behavior of biological counterparts in information processing is undoubtedly one of the remaining grand challenges in computing, which if resolved will improve energy efficiency and enrich functionality of existing electronics. In this talk we are going to present our most recent experimental results in the application of TiO2-based memristive crossbar array and floating-gate transistors in ANNs and in particular for pattern classification. Such results represent a fundamental step toward the demonstration of a full integration of these emerging technologies with the well-established CMOS circuits.

Gennadi Bersuker, Sematech (Gennadi.Bersuker88@gmail.com)

On the Way to Universal Non-volatile Memory: Structures, Mechanisms, Limitations

The accelerating drive to reduce power consumption and increase density/speed have reaffirmed focus toward further development of non-volatile memories technologies, in particular, resistance switching random access memory (RRAM), which structures are expected to satisfy a variety of requirements for advanced memory systems. In this talk, we discuss critical structural features and operating conditions controlling switching mechanisms in a promising group of metal oxide-based RRAM devices exhibiting a unique attribute of area independent electrical properties, which provides an ultimate device scaling advantage down to a few nm.

Shaun Kumar, GLOBALFOUNDRIES (shaun.kmr@gmail.com)

FinFET SRAM Design Challenges

Utilization of FinFET technology in SRAM provides several benefits over planar bulk devices for improved performance at lower leakage through vertical channel integration, as well as a local variation reduction through increase channel control of the gate. However, quantization of the channel width presents several unique challenges compared to planar devices. Two common FinFET SRAM cell types are discussed which suffer from reduced access disturb margin and write margin. Various assist methodologies, and their respective trade-offs, are discussed that can be employed to ameliorate the access disturb margin and write margin impact from the quantization of channel width.

Track M: Microelectronics & Nanotechnology

Session M1: Nanoelectronics

Carlo Reita, CEA (carlo.reita@cea.fr)

with S. Barraud

Dual-Channel CMOS Nanowire Device Co-Integration with Si NFET and Strained-SiGe PFET

Nanowire (NW) transistors are today widely recognized as a promising solution to pursue the Moore's law beyond FinFET and Fully-Depleted Silicon-On-Insulator (SOI) CMOS technologies. In this paper, we will report the successful co-integration of hybrid strained SiGe-channel (cSiGe) p-FETs and Si-channel n-FETs NW CMOS devices that outperform state-of-the-art SOI nanowires. In view of optimizing the performance of this hybrid CMOS technology it becomes necessary to study the advantages of uniaxial strain in deeply scaled n-FET NWs. The p-FET performance penalty expected from smaller hole mobility induced by the tensile strain will also examined and we will show the results of a systematic investigation of the strain-induced performance of high-κ/metal-gate π-Gate nanowire FETs fabricated by lateral strain relaxation of strained-SOI substrates.

Francis Balestra, MINATEC (balestra@minatec.grenoble-inp.fr)

Nanowires for Ultimate CMOS and Small Slope Switches

Nanowires are very promising for ultimate CMOS and Memories, Nanosensing and Energy Harvesting applications needed for future high performance terascale integration and autonomous nanosystems. For instance, Gate-All-Around Nanowire FETs could be considered as ultimate nanodevices with the best control of parasitic effects. They are also very interesting for developing small slope switches, using alternative source/drain/channel materials, for ultralow power circuits. This presentation will address the main challenges and possible solutions to continue Moore's law, the integration of many boosters (novel materials, strain, heterostructures, quantum wells...), needed for the development of future CMOS and Beyond-CMOS, with a particular focus on TFETs.

Takeshi Saito, National Institute of Advanced Industrial Science and Technology (takeshi-saito@aist.go.jp)

Structure-dependent Performances in the Electronic Applications of Single-wall Carbon Nanotubes

Extremely low variability with excellent device performances has been demonstrated in the array of single-wall carbon nanotube thin film transistors (SWCNT-TFTs) fabricated by using a semiconducting ink of short SWCNTs with an average length of 340 nm; the field-effect mobility of 3.9 - 0.45 cm2V(-1)s(-1), on/off ratio from 10^5 to 10^6, and small hysteresis of 0.5 V. It has been concluded that the shortening of SWCNTs in the dispersing process using the Brij 700 surfactant contributes to the observed uniformity of performance among the devices.

Michihisa Yamamoto, University of Tokyo (yamamoto@ap.t.u-tokyo.ac.jp)

Generation and Detection of Pure Valley Current in Bilayer Graphene

Valley is a useful degree of freedom for non-dissipative electronics since valley current does not accompany electronic current. We use dual-gated bilayer graphene in the Hall bar geometry to electrically control broken inversion symmetry or Berry curvature as well as the carrier density to generate and detect the pure valley current. A large nonlocal resistance and the scaling relation between the nonlocal resistance and the local resistivity allow us to prove existence of the valley Hall effect in the insulating regime where the local resistivity increases with lowering the temperature.

Anders Blom, QuantumWise A/S (anders.blom@quantumwise.com)

First-principles Simulations of Semiconductor Materials and Devices - Recent Progress and Remaining Challenges

QuantumWise is developing a modern, commercial software package for atomic-scale simulations which puts state-of-the-art methods in the hands of industrial R&D divisions. This enables companies to systematically investigate new materials—both exotic options like graphene and more common options like III-V random alloys—and to model novel architectures like tunnel FETs, nanowires, or spintronic device. A strong focus is placed on ease of use and the package encompasses a wide range of methods to enable users to address problems related to electronic, optical, mechanical and thermal properties.

Ricardo Reis, Universidade Federal do Rio Grande do Sul (reis@inf.ufrgs.br)

Physical Design Automation of Transistor Networks

Integrated Circuits implemented with traditional Standard Cell approaches use much more transistors than it is needed due to the reduced numbers of functions available in most of standard cell libraries. This also means more area, higher delays and more power consumption, mainly more static power consumption. An important way to reduce power consumption is to reduce the amount of transistors in a circuit, as leakage power is proportional to the number of transistors. It is shown a physical design approach to reduce the amount of transistors needed to perform a task. It is proposed an EDA tool set to automatically do the physical design of any transistor network. A standard cell library has also a limited number of sizings. The talk is target in optimization methods to reduce the amount of transistors of a circuit. The method allows the automatic generation of the layout of any possible logical function or transistor network, with any transistor sizing. It is included comparisons with solutions using the traditional standard cell methodology.

Session M2: Nanoelectronics

Garrett S. Rose, University of Tennessee, Knoxville (garose@utk.edu)

Assessing the Security Strengths and Vulnerabilities of Emerging Nanoelectronic Computing Systems

As a case study for security in emerging nanoelectronic computing, this talk will focus on memristor based systems. Given their low power operation and small footprint, memristors have emerged as excellent candidates for future memory and logic. However, the non-volatility of memristors presents certain security challenges whereby sensitive data may be vulnerable. At the same time, memristors also show promise for effective security primitives such as physical unclonable functions and random number generators. In this talk we will consider the security pros and cons of nanoelectronic systems and also discuss design techniques that best balance security concerns with performance needs.

Marc Riedel, University of Minnesota (mriedel@umn.edu)

Probability as a State Variable for Nanoscale Computation

Sergey Frolov, Pittsburg University (frolovsm@gmail.com)

Quantum Computing Based on Semiconductor Nanowires

A quantum computer will have computational power beyond that of conventional computers, which can be exploited for solving important and complex problems, such as predicting the conformations of large biological molecules. Materials play a major role in this emerging technology, as they can enable sophisticated operations, such as control over single degrees of freedom and their quantum states, as well as preservation and coherent transfer of these states between distant nodes. Here we assess the potential of semiconductor nanowires grown from the bottom-up as a materials platform for a quantum computer. We review recent experiments in which small bandgap nanowires are used to manipulate single spins in quantum dots and experiments on Majorana fermions, which are quasiparticles relevant for topological quantum computing.

David Jamieson, University of Melbourne (d.jamieson@unimelb.edu.au)

Single Engineered Donor Atoms with Nuclear and Electron Spin Readout for Quantum Bits

We use ion implantation to insert phosphorus atoms into silicon for quantum computer technology based on potentially scalable engineered single donor atom devices. We engineer nano-scale silicon CMOS devices with a single 31P atom implanted compatible with our deterministic doping method. In natural silicon these devices have an electron coherence time exceeding 200 μs and nuclear spin coherence times for ionized donors of 60 ms. Longer times are demonstrated in enriched 28Si. This presentation describes our approach to take this technology to the next stage by building deterministic arrays of single atoms with the goal of 6 nm positioning precision.

Andrew Dzurak, University of New South Wales (A.Dzurak@unsw.edu.au)

Spin-based Quantum Computing in Silicon

Spin qubits in silicon are excellent candidates for scalable quantum information processing due to their long coherence times and the enormous investment in silicon CMOS technology. While our Australian effort in Si QC has largely focused on spin qubits based upon phosphorus dopant atoms implanted in Si, we are also exploring spin qubits based on single electrons confined in SiMOS quantum dots. Such qubits can have long spin lifetimes T1 = 2 s, while electric field tuning of the conduction-band valley splitting removes problems due to spin-valley mixing. In isotopically enriched Si-28 these SiMOS qubits have a control fidelity of 99.6%, consistent with that required for fault-tolerant QC. By gate voltage tuning the electron g*-factor, the ESR operation frequency can be Stark shifted by > 10 MHz, allowing individual addressability of many qubits. Most recently we have coupled two SiMOS qubits to realize CNOT gates for which over 100 two-qubit gates can be performed within a two-qubit coherence time of 8 us. I will conclude by discussing the prospects of scalability of this technology using traditional CMOS manufacturing.

COFFEE BREAK (GEORGIA FOYER)

****

Sven Rogge, University of New South Wales (s.rogge@unsw.edu.au)

Donor Based Quantum Simulation

Quantum electronics exploiting coherent states of dopants in silicon requires interactions between states and interfaces, and inter-dopant coupling by exchange interactions. We have developed a low temperature STM scheme for spatially resolved single-electron transport in a device-like environment, providing the first wave-function measurements of single donors and exchange-coupled acceptors in silicon. For single donors, we directly observed valley quantum interference due to linear superpositions of the valleys. For exchange-coupled acceptors, we measured the singlet-triplet splitting, and extracted information about the 2-body wavefunction which allows for a hardware implementation of a quantum simulation.

Shaloo Rakheja, New York University (shaloo@mit.edu)

Quasi-ballistic Transport in Alternative Channel Material Transistors

In this talk, I will cover the fundamentals of quasi-ballistic (QB) transport in ultra-scaled transistors for advanced process technology nodes. First, the virtual-source (VS) model that provides a simple, physical description of transistors operating in the QB regime will be presented. This will be followed with experimental calibration of the model with DC I-V measurements of short-channel ETSOI and InGaAs HEMT devices. The VS model will be explained in context of the Lundstrom model allowing us to extract important technology parameters such as unidirectional thermal velocity, critical length for backscattering, and mean free path from the channel-length dependent input parameters (virtual source velocity and apparent mobility) in the VS model.

Jean-Christophe Blancon, Los Alamos National Laboratory (blanconjc@gmail.com)

Phase-engineering Low Resistance Contacts to Atomic Layered Transition Metal Di-chalcogenides for Optoelectronic Applications

The realization of viable electronics and opto-electronics technology based on new semiconductors requires low-resistance contacts for achieving practical drive currents and high-efficiency devices with low-power consumption. Ultrathin molybdenum disulphide (MoS_2) has emerged as an interesting new semiconductor because of its finite band gap and absence of dangling bonds; however transport performances of ultra-thin MoS_2 base field effect transistors (FETs) are plagued by relatively large contact resistance leading to Schottky behavior limited transport.

Here, we demonstrate significantly improved electronic performances after 1T phase transformation of MoS_2 at the source and drain electrodes in FETs. This phase engineering of the material decreases the contact resistance to record values (from 200 - 300 Ω-μm). Besides we investigate the local optoelectronics properties of MoS_2 channels with and without phase engineering transformation and confirm that a reduction of native Schottky barriers in a 1T device enhances the responsively by 40 times, a crucial parameter in achieving high-performance optoelectronic devices. Our results provide a new strategy based on phase engineering for achieving low resistance contacts and reproducible performance of FETs based on ultrathin MoS_2 and more generally to transition metal dichalcogenides.

Session M3: Nanotechnology

Shalom Wind, Columbia University (sw2128@columbia.edu)

Directed Biomolecular Assembly of Functional Nanostructures

Solution-based and high temperature synthesis of many types of functional nanomaterials limit their use in integrated systems on solid substrates. This talk will present new strategies for directing the precision assembly of metallic nanoparticles, quantum dots and carbon nanotubes in simple and complex topologies by combining high resolution nanolithography with biomolecular recognition using selected proteins and DNA. Also presented is the formation of hybrid heteronanostructures comprising functional nanomaterials with various molecular moieties having different structure and function. These approaches represent an important step toward the creation of hybrid, complex systems with functionalities that transcend those of their individual constituents.

Sasa Ristic, McGill University (sasa.ristic@mcgill.ca)

AFM Based Nanolithography

McGill Institute for Advanced Materials has recently purchased the world's first AFM-based nanolithography prototyping tool, "NanoFrazor Explore," made by SwissLitho AG. NanoFrazor uses an electrostatically actuated silicon cantilever with a sharp tip to sublimate polymer resist with microsecond-scale heat pulses. Holes can be made in the resist with better than 10nm of lateral and about 1nm of vertical resolution, enabling 3D nano patterns that can be read, and potentially corrected in-situ, before transferred into the underlying substrate. We demonstrate the technology by fabricating fiber-to-chip optical grating couplers based on regular gratings as well as highly directional, low-reflection multilevel-blazed gratings.

Aaron Lewis, Nanonics (aaron@nanonics.co.il)

Functional Nanofabrication and Imaging with a Multiprobe Workstation and Nanotoolkit

Our development of multiprobe scanned probe microscopy will be described. These tools have already been applied to a variety of interesting new horizons including plasmonic optical nanopump probe, single walled carbon nanotube & graphene electronics, ultra high resolution fluorescence, thermal nanodiffusion etc. These new horizons in nanotechnology being addressed with this SPM based probe station design which includes a NanoToolKitTM of probes suitable for multiprobe operation also permits on-line optical and electron optical characterization including Raman scattering to monitor chemical alterations occuring during the probed functional process.

Byron Gates, Simon Fraser University (bgates@sfu.ca)
with M.C.P. Wang

Polystyrene Core-Shell Copper Selenide Nanowires Prepared by Simultaneous Chemical Transformation and Surface-Initiated Polymerization

We developed a method to prepare copper selenide nanowires through solution-phase chemical transformation that simultaneously encapsulates these nanostructures with a thin polystyrene coating to improve their chemical stability in corrosive environments. A simple process was sought for preparing copper selenide and similar nanostructures while also adding a protective coating without additional processing steps or processing steps or processing steps or processing steps or significantly increasing the overall dimensions of nanostructures. Selenium templates are transformed through electrochemical processes simultaneous to a surface initiated atom transfer radical polymerization (SIlymerization (SI -ATRP) reaction. Composition of the encapsulated nanowires is investigated by a number of spectroscopy and microscopy techniques.

Jean-Pierre Cloarec, Centre National de la Recherche Scientifique (jcloarec@ec-lyon.fr)

##

Addressing Large Sets of Multifunctional Particles onto Pre-defined Nano-sites Using Selective Surface Chemistries

Finding simple methods for the precise self-assembly of colloids onto different regions of a patterned surface is a major current issue in nanofabrication. Solving this issue could bridge the gap between top-down fabrication processes such as lithography which allows the creation of large arrays of well-defined nanostructures on a surface (e.g. nanoantennas) and bottom-up built nano-objects. In our work, latex nanoparticles were precisely located onto the gold regions of micro and nanopatterned gold/silica substrates through selective surface chemical functionalizations. This strategy allowed the trapping of nanoparticles onto the gold nanostructures with very little non-specific adsorption onto the surrounding silica.

Session M4: MEMs and NEMs

Songbin Gong, University of Illinois at Urbana-Champaign (songbin@illinois.edu)

Piezoelectric and Electrostatic MEMS Devices for Future Reconfigurable RF Front Ends

The growth in functionalities of modern wireless communication, coupled with the Defense's need for a programmable RF framework, is drawing great interest in developing reconfigurable radio frequency (RF) systems. Having access to programmable subsystems will revolutionize the RF design space, improve cost and yields, and enable in-field system adaptation to RF signals and novel RF transceiver architectures. Micro-scale RF passives, namely micro-electro-mechanical systems (MEMS) resonators, are ideal candidates for the implementation of such reconfigurable RF platforms.

This talk will present some the most promising RF MEMS technologies that I have developed for intelligent and efficient utilization of the RF spectrum, namely the Lithium Niobate (LN) laterally vibrating resonators (LVRs). The LN LVRs, which have record-breaking figure of merit (FoM) as well as the capability of covering multiple frequencies on a single chip, will be described as the key building blocks for programmable filtering. I will describe the design and fabrication challenges that have been addressed, and the subsequently demonstrated first LN LVRs for chip-scale reconfigurable filtering applications. The design and development of these RF micro-systems will be discussed with a special focus on how these devices can be assembled monolithically to enable unprecedented adaptive filtering capabilities on system level.

I will also show that in short term (2~3 years), such system-level capabilities promise the highly sought-after multi-band (~20) multiplexing solutions for the existing standards such as LTE, CDMA, WiFi, and GPS; and in long term, they can break down the design boundaries imposed by the conventional device/hardware infrastructure and favorably respond to the growing wireless bandwidth demand fueled by the uprising mobile paradigms such as cloud computing and internet of everything.

Douglas Buchanan, University of Manitoba (Douglas.Buchanan@umanitoba.ca)

Advanced MEMS Ultrasonic Transducers for Non-Destructive Testing and Imaging

The fabrication and operation of a novel capacitive micromachined ultrasonic transducer will be discussed. This device employs a stack of two deflectable membranes. Both moving membranes deflect simultaneously when biased. This results in a smaller effective cavity height when compared to conventional capacitive transducers. Electrical and physical measurements have been compared with analytical models and conventional, single membrane devices on the same chip. A larger membrane deflection and smaller effective cavity height are achieved with the double membrane structures which can enhance the transducer acoustic power generation and sensitivity.

Vamsy Chodavarapu, McGill University (vamsy.chodavarapu@mcgill.ca)

Ultra-high Quality Factor CMOS-MEMS Solutions for Timing and Synchronization

The advent of modern portable electronic devices has imposed stringent requirements on the frequency references they use. Existing technologies such as quartz crystal oscillators are unable to meet these demands in terms of cost, size and power consumption. RF MEMS oscillators have attracted significant interest in the past decade as they promise to overcome the limitations of quartz technology. However, their commercialization has been a rather slow process. In this talk, we will discuss our progress in overcoming the key challenges of the technology in developing high performance all-silicon resonators in a commercial MEMS process. We will describe our development of die-level vacuum encapsulated devices that are characterized by a high electrostatic efficiency and ultra-high quality factors that are on the order of 2.4 Million.

Edmond Cretu, University of British Columbia (edmondc@ece.ubc.ca)

Mode Localization Operation in Weakly-coupled Resonators: A Route Towards High-sensitivity Sensing in Electronics and MEMS Transducers

Mina Rais-Zadeh, University of Michigan (minar@umich.edu)

GaN Resonators and AlGaN/GaN Resonant Body Transistors

For decades, the electronics industry has benefited mainly from integration and scaling of transistors. To address more challenging applications, it is required to monolithically or heterogeneously integrate robust multi-functional devices and material systems on the same chip or in the same package. A GaN-on-silicon platform can address this need: GaN offers a number of excellent mechanical properties such as strong piezoelectric and pyroelectric effects, and high mechanical and chemical stability, while silicon is a perfect interposer. In this talk, I will present a new class of devices using a GaN-on-silicon platform, and discuss their application in timing and integrated sensing.

COFFEE BREAK (GEORGIA FOYER)

****

Kirill Poletkin, Universität Freiburg (k.poletkin@gmail.com)

Levitation Technology in Micro-machined Sensors and Actuators

The levitation technology is an indisputable solution to eliminate completely mechanical attachment, consequently the friction, between stationary and moving parts of a micro-sensor and micro-actuator. Besides, for a micro-actuator the levitation provides the extension of a motion range of its moving parts and reduces wasted energy significantly. Although different technologies can be used to provide the levitation, however electro-magnetic levitation is practicable and has already attracted a great deal of attention from researchers. In this talk the overview of electro-magnetic levitation technology in micro-sensors and -actuators and its perspective are discussed.

Philip Feng, Case Western Reserve University (philip.feng@case.edu)

Two-Dimensional (2D) Nanoelectromechanical Systems (NEMS) in Atomically-Thin Semiconducting Crystals

This talk introduces 2D NEMS in atomic-layer crystals. Beyond graphene, the well-known herald of 2D crystals, semiconducting 2D crystals with tunable bandgaps have emerged, such as atomic layers of transition metal di-chalcogenides (TMDCs) and black phosphorus. These 2D crystals possess a number of interesting electrical, optical, and mechanical properties, and are enabling a new class of NEMS. I will describe recent experiments on various radio-frequency graphene, MoS_2 and other 2D NEMS, with outstanding tunability. Through sensitive optical and electronic measurements, in combination with modeling, we quantify the performance and potential of these 2D NEMS toward ultralow-power signal processing and sensing.

Matteo Rinaldi, Northeastern University (rinaldi@ece.neu.edu)

Hybrid Piezoelectric NEMS Resonant Nano Plates for Advanced Sensing and Wireless Communications

Sensors are nowadays found in a wide variety of applications, such as smart mobile devices, automotive, healthcare and environmental monitoring. The recent advancements in terms of sensor miniaturization, low power consumption and low cost allow envisioning a new era for sensing in which the data collected from multiple individual smart sensor systems are combined to get information about the environment that is more accurate and reliable than the individual sensor data. By leveraging such sensor fusion it will be possible to acquire complete and accurate information about the context in which human beings live, which has huge potential for the development of the Internet of Things (IoT) in which physical and virtual objects are linked through the exploitation of sensing and communication capabilities with the intent of making life simpler and more efficient for human beings.

This trend towards sensor fusion has dramatically increased the demand of new technology platforms, capable of delivering multiple sensing and wireless communication functionalities in a small foot print. In this context, Micro- and Nanoelectromechanical systems (MEMS/NEMS) technologies can have a tremendous impact since they can be used for the implementation of high performance sensors and wireless communication devices with reduced form factor and Integrated Circuit (IC) integration capability.

This work presents a new class of Aluminum Nitride (AlN) piezoelectric nano-plate NEMS resonant devices that can address some of the most important challenges in the areas of physical, chemical and biological detection and can be simultaneously used to synthesize high-Q reconfigurable and adaptive radio frequency (RF) resonant devices. By taking advantage of the extraordinary transduction properties of AlN combined with the unique physical, optical and electrical properties of advanced materials such as graphene, photonic metamaterials, phase change materials and magnetic materials, multiple and advanced sensing and RF communication functionalities are implemented in a small footprint. Particular attention is dedicated to the key attributes of such piezoelectric MEMS/NEMS devices in realizing intrinsically switchable and reconfigurable RF MEMS components, high performance gravimetric chemical sensors, ultra-fast and high resolution un-cooled IR/THz detectors and ultra-miniaturized and low power magnetoelectric sensors.

Feng Zhao, Washington State University (feng.zhao@wsu.edu)

Single-Crystalline 4H-SiC MEMS: Fabrication and Characterization

Fabrication and characterization of single crystalline 4H-SiC based MEMS devices will be reported. As a wide bandgap semiconductor with excellent material properties, 4H-SiC is a desirable platform for MEMS operation in harsh environments, such as high temperature, extreme pressure, chemical, radiation, biological, etc. With both p-n and n-p-n homoexpitaxial 4H-SiC structures, applications of MEMS devices in electrostatically actuated sensors and monolithic integration with electronic devices and circuits are attainable.

Session M5: Microelectronics

Ramgopal Rao, Indian Institute of Technology Bombay (rrao@ee.iitb.ac.in)

Bottom-up Meets Top-down: An Integrated Approach for Future Nanoscale CMOS

Conventional CMOS technologies employ top-down fabrication methodologies for high volume manufacturing. However, as the CMOS technologies are scaled down, there are many challenges owing to the variability, reliability and power issues. Some of these issues in nano-scale CMOS technologies can be better addressed by employing a host of bottom up nanotechnology approaches through innovative process integration strategies. We will show in this talk how Self Assembled Monolayers (SAMs) can be used as diffusion barriers for Cu interconnects, for work-function tuning applications in metal gate technologies and also as dopant sources in Finfets. With the help of Applied Materials, a selective SAM formation has also been recently demonstrated using a vapor phase approach, which paves way for integration of these SAM processes in future advanced CMOS technology nodes.

Lan Wei, Massachusetts Institute of Technology (lanwei@ieee.org)

Technology Assessment Based on Device and Circuit Interactive Design and Optimization

Continued progress in nanoelectronics necessitates a holistic view across the boundaries of device, circuit and system. Device engineering and circuit design must be interactively explored targeting improvement at circuit and system level. In this talk, new benchmarking methodologies for an apples-to-apples comparison among different device structures is proposed based on optimized circuit- and system- level performance. The methodology links the device-level behaviors and circuit-level performance and energy efficiency, and further extended into an optimizing tool for device/circuit interactive design. In addition to speed and energy, yield is also incorporated into the co-optimization framework, paving the path for reliability-aware circuit design.

Patrick Mercier, University of California, San Diego (pmercier@ucsd.edu)

Next-generation Switched-capacitor DC/DC Converter Topologies

Mutsumi Kimura, Ryukoku University (mutsu@rins.ryukoku.ac.jp)

Novel Architecture on Neural Network of Device Level

Artificial neural networks are promising information processors with many advantages, including self-teaching and parallel distributed computing. However, conventional ones consist of extremely intricate circuits to guarantee accurate behaviors. We demonstrate a self-organized electronic device using thin-film transistors. First, we formed a "neuron" from only eight transistors and reduced a "synapse" to only one transistor by employing characteristic degradations to adjust synaptic connection strengths. Second, we classified synapses into "concordant" and "discordant" synapses and composed a local interconnective network optimized for integrated circuits. Finally, we confirmed that the device could work and learn multiple logical operations, including AND, OR, and XOR.

Kirk Bevan, McGill University (kirk.bevan@mcgill.ca)

Can We Achieve Atomic Scale Imaging Resolution of the Voltage Drop in 2D CMOS Materials?

Advanced measurement techniques possessing nanoscale and atomic scale resolution have played a pivotal role in the development of electronic devices over the past decade. Amongst all such characterization tools, scanning tunneling potentiometry (STP) provides indispensable insight into the nature of the potential drop within such devices. Though this is now routinely achievable with scanning tunneling microscopy (STM), the conclusive observation of atomic scale features in STP imaging has remained elusive. Motivated by the desire to better chart this unknown, we present a theoretical first-principles study of atomic scale STP imaging at a graphene grain boundary.

Session M6: Microelectronics

Stefano Esposito, Politecnico di Torino (stefano.esposito@polito.it)

with M. Violante

COTS-Based High-Performance Computing for Space Applications

The peculiarity of the radiation environment in which space systems operate makes them a special class of computing systems. The environment interferes with the operations of electronic systems inducing either permanent or transient misbehaviors. When designing a space system, special care must be given to the radiation effects, mainly because a space system cannot be easily reached for maintenance, as a result high reliability is a key requirement for this kind of systems.

The typical solution to obtain the needed reliability has been the use of special components which are in some cases specifically designed to cope with the radiation effects as in the so-called Radiation Hardened (rad-hard) components, while in other cases are normal components carefully selected and inserted in a suitable packaging to create the so-called Radiation Tolerant (rad-tolerant) components.

However, more recent space systems' computation requirements are greatly increasing and the rad-hard and rad-tolerant components cannot cope with such requirements. This is due to the fact that to be certified as space-graded such components must pass through a very long and expensive verification phase and as such they are typically not at the very edge of the current technology.

Space systems up to now uses specific solutions like LEON3 which grants needed reliability in the space radiation environments, although LEON3 cannot grant the needed performances. To achieve such performances, use of commercial off-the-shelf components has been proposed. Such components can achieve a high reliability through Software Implemented Fault Tolerance (SIFT), but the validation of such strategies can be difficult and costly.

This work focuses on the application of techniques in a space system including COTS components and on the validation of such techniques.
A SIFT strategy based on time redundancy and information redundancy has been implemented. Time redundancy has been exploited by repeating the execution, in a Virtual Duplex System configuration. Information redundancy has been exploited both by duplicating working data and by encoding memory with an error correcting code.

To validate space graded systems an irradiation chamber is usually needed to study system's behavior when hit by high energy particles such the ones it will encounter during operations. However such chambers are few and rather costly.

To validate the strategy implemented in this work, a fault injection approach has been used. The fault injection approach does not need a radiation chamber, as such it is less expensive. The fault injection system implemented for this work uses an instance of the system to validate and injects faults directly into the system to study its reactions.

The fault injections campaign emulated the effects of radiations by injecting SEUs in the processor registers.

Results from the fault injection campaigns, shows that the implemented SIFT strategy is capable of making the target system immune to SEU, thus making it suitable for space operations.

Stephan Breitkreutz-von Gamm, Technische Universität München (stephan.breitkreutz@tum.de)

3D-integrated Magnetic Computing: Current State and Future Challenges of Perpendicular Nanomagnetic Logic
Perpendicular Nanomagnetic Logic (pNML) is an emerging beyond-CMOS technology employing the 3-dimensional field-coupling of bistable nanomagnets to perform logic operations. Nonvolatility, majority logic, high density integration, low power computing, zero leakage and CMOS compatibility are key features of pNML.
In this talk, the current state of 3D-integrated pNML circuits is shown and reviewed in terms of physical measures (area, power, speed) and operational reliability. Signal routing concepts and suitable architectures for complex pNML systems are discussed. Future challenges are outlined based on simulations, which explore the performance of pNML in terms of scaling and material improvements.

Sandipan Pramanik, University of Alberta (spramani@ualberta.ca)

Giant Current-Perpendicular-to-Plane (CPP) Magnetoresistance Effect in Multilayer Graphene Stacks
Magnetoresistance in graphitic systems has drawn significant attention in recent years. In this work we consider multilayer graphene as grown on nickel and study c-axis charge transport when the magnetic field is applied normal to the graphene plane. We show that the electrical resistance measured across the graphene stack can be reduced by two orders of magnitude by applying a relatively small magnetic field of few kilogauss normal to the layer plane. This feature persists even at room temperature and is far stronger than any other magnetoresistance effect reported to date for comparable temperature and field conditions. This effect can be qualitatively explained within the framework of "interlayer magnetoresistance" (ILMR). Existence of such effect makes multilayer graphene an attractive platform for magnetic field sensing, data storage and exploration of fundamental insights into graphene physics.

Edward T. Yu, University of Texas (ety@ece.utexas.edu)

Epitaxial Oxides on Silicon for Resistive Switching Memories with Controllable Conductance Quantization
Resistive switching in single crystal anatase TiO2 on SrTiO3/Si (001) is characterized and analyzed. Although switching occurs via a valence-change mechanism involving motion of oxygen vacancies, electrical characteristics highly reminiscent of electrochemical metallization memories (low leakage current, highly linear current-voltage characteristics in the low-resistance state, and on/off current ratio up to ~107) are observed. We also demonstrate quantized conductance in the low-resistance state, with the number of conducting ballistic channels being controlled via compliance currents applied during the SET process. Conductances corresponding to 1 to 4 conductance channels, controllable to within a single quantum of conductance, are observed.

Alberto Riminucci, Consiglio Nazionale delle Ricerche Bologna (alberto.riminucci@gmail.com)

with M. Prezioso, M. Calbucci, P. Graziosi, R. Cecchini, I. Bergenti and V.A. Dediu
Magnetically Enhanced Memristor: Opportunities and Challenges

Information and communication technology (ICT) is calling for solutions that achieve lower power consumption, further miniaturization (Moore's law) and multifunctionality. This requires the development of new device concepts and new materials. A fertile approach to meet these demands is spintronics, that is the introduction of the spin degree of freedom into electronic devices. Spintronics already lead to a revolution in information storage (e.g. giant magnetoresistance (GMR) readheads) in the last 20 years. Nowadays, the challenge is to bring spintronics also into devices dedicated to logic operations, data communication and storage, and to do it within the same materials technology. One of the most promising approaches is the use of arrays of crossbar memristors capable of information processing and storage.

In this context the electric control of the magnetoresistance is one of the most promising approaches, allowing both further miniaturization and multifunctional operation of spintronic devices. We show that the magnetoresistance of organic devices can be controlled electrically by combining magnetic bistability(i.e. the spin valve effect) and resistive switching. In such devices the GMR effect can be turned on and off by a programming bias that sets the device in a low or a high resistance state respectively. The magnitude of the GMR depends on the history of the applied bias and can be controlled reversibly; we show that such devices operate as a magnetically enhanced memristors (MEM). MEMs can be used both in memory and in logic gate applications and can lead to new device concepts.

One of the greatest challenges to the understanding of the fundamental physics of these devices is the fact that the effect of spin precession, also known as the Hanle effect, is not observed. Spin precession in an incoherent transport regime should cause the loss of magnetoresistance when a magnetic field is applied perpendicular to the magnetization of the injecting electrodes. Such effect is not observed, and we discuss some reasons why it should be so.

COFFEE BREAK (GEORGIA FOYER)

****

Edwin Kan, Cornell University (kan@ece.cornell.edu)

Hardware-oriented Security: Unclonable ID and Unpredictable Randomness

John von Neumann pointed out that random numbers, and hence many related cryptographic protocols, need to be extracted from hardware instead of arithmetic procedures. To formulate security keys, random numbers and information hiding, low-level hardware-oriented security functions are mostly based on unclonable process variations and inherent random noises, both of which are minimized in normal logic functions but need to be reliably and efficiently extracted to execute the security protocol. We will present how to directly extract the unclonable but repeatable ID and the true random noises from a single Flash cell based on single-particle effects, instead of long timing chains or positive feedback metastability, which greatly enhances the layout efficiency and hence the overall throughput in bits per second as well as bits per joule. The single particle physics employed include the effect of one dopant atom in the transistor channel, one electron trapped at the interface, and one trap generation with the tunnel oxide.

We have benchmarked our security extraction algorithms experimentally to estimate the inherent entropy by separating contributions from various variation and noise sources, including: unclonable ID by discrete random dopant fluctuation (RDF), true random number generation (TRNG) by random telegraph noise (RTN), and covert data encoding (CDE) by cyclic endurance aging (CEA). These cell-level statistical variations and noises can be directly and reliably extracted from the standard NAND Flash circuit interface with no need of custom circuits, because the program and erase characteristics of Flash memories take much longer than common processor instruction cycles and can be readily measured. The security level, system entropy and unclonability have been mapped to the underlined physics with great reliability under harsh conditions such as reverse-engineering abrasion, cold attack and full-chip replay. As most mobile electronic systems contain Flash memory, the Flash-based security function provides a ready implementation for ubiquitous security.

Aurélie Thuaire, CEA (aurelie.thuaire@cea.fr)
##

Atomic Scale Device Packaging: Challenges and First Silicon Technology Developments

Atomic scale device packaging relies on very challenging specifications especially regarding surface preservation and interconnection. Electrical interconnections from the atomic-scale circuit to the macroscopic environment require the development of a specific connector chain with a fan-out ranging from the nanometer to a few hundreds of micrometers. Surface and circuit preservation require a hermetic packaging which must be reversible in order to be able to release the cap for special dedicated processes. The technological process flow for such a die realization will be presented as well as our recent silicon technology developments on the major elementary steps of the process flow.

Ken Brizel, ACAMP (kbrizel@acamp.ca)

Submicron Alignment and Welding for Optical Assemblies

Optoelectronic assembly differs from other microelectronic assembly due to the need for precision submicron optical alignment of single mode optical fiber to active photonic components. There are two major methods of achieving this alignment. First is the use of epoxy bonding with 6 axis alignment stages in a generally manual process. The second, often more robust process, is the use of an automated laser align and weld station. Laser welding has several advantages including shorter process time, less post bond shift of the optics and a more robust final assembly.

Olivier Pollet, CEA (olivier.pollet@cea.fr)

with S. Barnola, N. Posseme and P. Pimenta-Barros

CMOS Devices Beyond 10nm: An Upcoming Transition Era for the Etching/Stripping Engineer

To achieve gate electrostatics control meeting specifications for sub-10nm logic devices, implementation of a three-dimensional channel thoroughly or partially surrounded by the gate is an unavoidable step. Additionally higher mobility materials are required to build transistor channel complying with ON-state current expectations, such as germanium or compound semiconductor. Combining these changes in logic integration with ever-shrinking feature size brings up substantial challenges for etching and stripping. Indeed profile control, etch selectivity, pattern collapse-free processing become dramatically important when fabricating very high aspect ratio features, such as trigate, or 3D structures, such as silicon nanowires. Moreover uncommon properties of new channel materials such as low by-product volatility for Indium-based III-V compounds or high sensitivity to oxidation for Germanium alloys drives new etching and stripping developments towards quite different operating conditions than to-date usual processes. This presentation will focus on challenges, solutions and results achieved at Leti to address sub-10nm device manufacturing.

Track O: Optoelectronics & Photonics

Session O1: Photonics & Solar Energy

Bruno Matarese, Imperial College London (b.matarese13@imperial.ac.uk)

Organic Light Emitting Diodes (OLEDs) for Optical Stimulation of Neurons

The present talk will aim to show that the stimulation of neurons by OLEDs can offer advantages over conventional inorganic light sources for Optogenetics. Here, will be discussed the development of biocompatible organic light-emitting diodes (OLEDs) for incorporation into living tissues, particularly for controlled photo-stimulation of neurons. Additionally, the challenge to identify appropriate OLED materials and different device architecture suitable for the neurobiological constraints in terms of water and oxygen exposure will be examined. The field of applications from OLEDs will be presented and discussed at the conference.

Hatice Altug, École Polytechnique Fédérale de Lausanne (hatice.altug@epfl.ch)

with A. E. Cetin, B. Galarreta, M. Huang and D. Etezadi

Nanobiophotonics for Advanced Diagnostics and Biospectroscopy Applications

Highly sensitive biosensors are needed for early diagnostics and preventing epidemics. To understand disease origins and find effective treatments, detection platforms should also screen large variety of bio-chemicals and cellular processes at high-throughput. In developing countries, there is an urgent need for cost-effective, portable and easy-to-use diagnostics technologies. To address these needs in life sciences, biomedical and biotechnology applications, Dr. Altug's laboratory is developing bio-detection systems using nano-plasmonics and microfluidics. In her talk, she will introduce a high-throughput and label-free protein microarray technology with nearly one million sensor elements for reliable and quantitative detection of bio-chemicals.

Giovanni Fanchini, University of Western Ontario (gfanchin@uwo.ca)

with R. Bauld, A. Akbari-Sharbaf, A. Venter, M. Hesari and M. S. Workentin

Nanoplasmonic Organic Solar Cells Incorporating Molecular Gold Nanoclusters

Molecular gold nanoclusters (Aun) formed by n = 25, 38 or 144 atoms of gold possess unique properties and several advantages over more frequently used gold nanoparticles (Au-NPs). Whereas ensembles of Au-NPs are unavoidably affected by random distributions in diameters, shape and photo-physical properties, molecular gold nanoclusters exhibit a deterministic molecular structure and behavior.

In this talk, we will review the research carried out by our group in nanoplasmonic photovoltaic applications of Aun molecules protected by specific ligands. We demonstrate that aggregates of intact Aun molecules and perfectly spherical gold micro-globules can be obtained from in situ nucleation of Au25-, Au25+, neutral Au25 and Au144 molecules in polyimide at temperatures from 150°C to 450°C. The different charge state of Au25-, Au25+, neutral Au25 leads to very different magnetic properties of Au25 thin films, since Au25- is diamagnetic while neutral Au25 and Au25+ in the solid state are paramagnetic with spin state ½ and 1, respectively.

Using scanning near-field optical microscopy and UV-visible spectroscopy, we demonstrate the presence of strong plasmonic resonance, which opens up a variety of applications in low-cost organic optoelectronics and plasmonics for our fully solution-processed thin films. We prepared nanocomposite hole-blocking layers consisting of poly-3,4-ethylene-dioxythiophene:poly-styrene-sulfonate (PEDOT:PSS) thin films incorporating networks of Au144 molecular precursors that may enhance of up to 10% the photoconversion efficiency of bulk heterojunction organic solar cells. Varying the concentration of Au144 in the starting solution, different morphologies of gold nanostructures can be obtained on indium tin oxide, graphene and other substrates, from individual nanoparticles (Au NPs) to tessellated networks of gold nanostructures (Tess-Au144). Improvement in organic solar cell efficiencies relative to reference cells is demonstrated with Tess-Au144 embedded in PEDOT:PSS.

Maxime Darnon, Université de Sherbrooke (maxime.darnon@usherbrooke.ca)

Solar Cell with Gallium Phosphide / Silicon Heterojunction

Current amorphous silicon/crystalline silicon heterojunction solar cells are limited by optical losses in the front a-Si layers that limit the short circuit current. In this work, we propose to use a thin GaP layer to replace the front a-Si layers because of better properties in the UV region. We show with a non-optimized GaP deposition process that the better transparency in the UV range promises a gain of more than 1 mA.cm-2 but is counterbalanced by a loss of efficiency in the IR region associated principally with bulk minority carrier lifetime degradation during the surface preparation preceding GaP deposition.

Lionel Vayssieres, International Research Center for Renewable Energy (lionelv@xjtu.edu.cn)

Low Cost Nanodevices for Solar Water Splitting

Some of the latest advances in low cost nanodevices for solar water splitting at neutral pH, without sacrificial agents and at low bias will be presented. Visible-light active quantum-confined oxide heteronanostructures based on earth abundant large bandgap semiconductors engineered to efficiently separate electron-hole pairs and drive chemical reactions at their interface have been fabricated. Their optical, electronic structure, dimensionality effects and interfacial properties along with their efficiency have been thoroughly investigated at synchrotron radiation facilities as well as in our laboratories. The most promising structures will be revealed and discussed.

Session O2: Nanophotonics

Christoph Gadermaier, Jožef Stefan Institute (christoph.gadermaier@ijs.si)

From Photophysics to Optoelectronics of Layered Transition Metal Dichalcogenides

Kirill Bolotin, Vanderbilt University (kirill.bolotin@vanderbilt.edu)

Optoelectronics of 2D Materials

The focus of this talk is a complex interplay between electrons, excitons, and photons in two-dimensional semiconductors such as monolayer molybdenum disulfide (MoS2). First, I will discuss strain-induces changes in the bandgap and phonon spectra of MoS2. Second, I will demonstrate the use of photocurrent spectroscopy to study excitons in ultraclean suspended MoS2 specimens where substrate-induced screening is suppressed. Finally, I will demonstrate electrical modulation of the near-field energy transfer between MoS2 and a layer of chemically synthesized quantum dots.

Fabrice Vallee, Université de Lyon (fabrice.vallee@univ-lyon1.fr)
with A. Lombardi, E. Pertreux, A. Crut, P. Maioli and N. Del Fatti

Optics of Hybrid Metal-semiconductor Nano-systems

The optical properties of nano-hybrids formed by different materials are little studied, though they offer wide possibilities for developing novel plasmonic systems. It also raises fundamental questions on the coupling mechanisms between the forming materials. The ultrafast nonlinear responses of nano-hybrids were investigated in ensemble of Au-CdS nano-matchsticks and in single Ag@SiO2-Au hetero-dimers. The results show ultrafast electron transfer between Au and CdS, while a Fano effect is demonstrated in Ag@SiO2. In the latter, the possibility to selectively heat and/or detect each component opens the way to investigate nanoscale energy transfer between two different nano-object separated by a few nanometers.

Lih Y. Lin, University of Washington (lylin@uw.edu)
with P. Jing, J. Wu and E. Keeler

Trapping, Sensing and Photostimulation of Cells through Nanostructures

We discuss enhanced optical trapping and photostimulation of live cells through two nanostructures: quantum dots (QD) and photonic crystals (PhC). Understanding how our brain works ultimately requires new technologies to stimulate and record neurons non-invasively. While optogenetics has shown impressive results in photostimulating neurons, its sensitivity is low and optical fibers are typically required to deliver light. We have demonstrated photostimulation of neurons using semiconductor QDs with sensitivity of 0.0036 mW/mm2, significantly higher than state-of-art optogenetics approach using red-shifted cruxhalorhodopsin (0.756 mW/mm2). We also discuss the potential of non-toxic Si QDs for this application. Through interaction with PhC, we have achieved optical trapping of nanoparticles with intensity ~2 orders of magnitude lower than conventional optical tweezers. We will discuss our work in viability study of live cells using this technology, and the potential of sensing and monitoring cell mass by integrating it with MEMS resonators.

Chennupati Jagadish, Australian National University (Chennupati.jagadish@anu.edu.au)

III-V Semiconductor Nanowires for Optoelectronics Applications

Semiconductors have played an important role in the development of information and communications technology, solar cells, solid state lighting. Nanowires are considered as building blocks for the next generation electronics and optoelectronics. In this talk, I will introduce the importance of nanowires and their potential applications and discuss about how these nanowires can be synthesized and how the shape, size and composition of the nanowires influence their structural, electronic and optical properties. I will present results on axial and radial heterostructures and how one can engineer the optical properties to obtain high performance optoelectronic devices such as lasers, solar cells and THz detectors. Future prospects of the semiconductor nanowires will be discussed.

COFFEE BREAK (GEORGIA FOYER)

****

Douglas M. Gill, IBM (dmgill@us.ibm.com)

CMOS Compatible Monolithic Traveling Wave Electro-Optic Modulators

A high level overview of the IBM CMOS Integrated Nano Photonic (CINP) technology and its fabrication process flow will be given. A monolithic Mach-Zehnder modulator based transmitter with integrated CMOS RF-drive that gives an OMA link sensitivity comparable to a 25 Gb/s commercial reference transmitter will be highlighted. In addition, various CINP traveling wave transmitters will be discussed including Mach-Zehnder modulators with integrated terminating resistor networks, de-coupling capacitors, and inductors, which demonstrate examples of how monolithic CINP technology platforms offer many options for design optimization.

Stephane Albon Boubanga Tombet, Tohoku University (stephanealbon@hotmail.com)

Emission and Detection of THz Radiation Using Graphene and III-V Semiconductor Heterostructures

Recent advances in emission and detection of terahertz (THz) radiation using graphene and III-V semiconductors are presented. First topic focuses on graphene, a monolayer carbon-atomic honeycomb lattice crystal, exhibiting peculiar carrier transport and optoelectronic/plasmonic properties owing to massless and gapless energy spectrum. Theoretical and experimental studies toward the creation of graphene terahertz injection lasers are described. Second the two-dimensional plasmons in III-V semiconductor heterostructures are discussed to demonstrate intense broadband emission and ultrahigh-sensitive detection of THz radiation. The device structure is based on a high-electron mobility transistor and incorporates the author's original asymmetrically interdigitated dual-grating gates.

Connie Chang-Hasnain, University of California, Berkeley (cch@berkeley.edu)

Nanophotonics on Silicon

Maiken Mikkelsen, Duke University (m.mikkelsen@duke.edu)

Radiative Decay Engineering Using Plasmonic Nanostructures

We demonstrate radiative decay engineering of dye molecules embedded in plasmonic nanoantennas with sub-10-nm gap sizes. The nanoantennas consist of colloidally synthesized silver nanocubes separated from a metal film by a ~5 nm dielectric spacer layer with embedded fluorophores. For antennas resonant with the excitation wavelength we observe fluorescence enhancements exceeding a factor of 30,000 [Rose et al., Nano Letters 14, 4797 (2014)]. Next, we probe the nanoscale photonic environment of the embedded emitters enabling the demonstration of spontaneous emission rate enhancements exceeding 1,000 while maintaining high quantum efficiency and directional emission [Akselrod et al., Nature Photonics 8, 835 (2014)].

Session O3: Photonics

Jean-François Pratte, Université de Sherbrooke (Jean-Francois.Pratte@USherbrooke.ca)

3D Single Photon Counting Modules: Status and Future Work

Optoelectronics systems have become the ubiquitous backbone of a multitude of applications, from smartphones and tablets to medical imaging instruments and telecommunication systems. They are in constant evolution, with performance improvement targets such as increasing the number of pixels, photodetection efficiency, timing measurement accuracy and the integration of smart microsystems for signal processing. This talk is about the realisation of 3D Single Photon Counting Modules, where an array of photodetectors is integrated on top of the required front-end microelectronic circuits. The microsystem features single photon detection capability, the ability to timestamp photons with 10ps accuracy and advanced digital signal processing.

Dayan Ban, University of Waterloo (dban@uwaterloo.ca)

Scanning Voltage Microscopy Measurement of Lasing Photonic Devices

The operation and performance of active photonic devices are governed by inner workings such as electric potential distribution and dynamic charge carrier distribution, which can conventionally be calculated from theoretical modeling but rarely be measured directly from experiments. The experimental characterization of active photonic devices is mainly focused on input/output behaviors or static structural information. In the former case, the devices are in operation but no nanoscopic information can be obtained. In the latter case, nanoscopic structural information can be obtained by using scanning electron microscopy and transmission electron microscopy but the devices under inspection are not in their working condition.

Scanning voltage microscopy (SVM) is a novel and enabling tool to quantitatively probe internal voltage distribution and carrier distribution at nanometer scales. In this talk, I will present our recent experimental study by applying scanning voltage microscopy to two representative photonic devices: terahertz (THz) quantum cascade laser (QCL) and interband cascade laser (ICL). Non-uniform electric field in the active region of a lasing THz QCL is observed and systematically investigated for the first time. In the ICL case, our SVM results clearly confirm the accumulation and spatial segregation of holes and electrons in the quantum-well active region. The charge carrier densities in the active region are measured as a function of device bias.

Werner Hofmann, Technischen Universität Berlin (Werner.Hofmann@tu-Berlin.de)

Nanostructured VCSELs for Optical Interconnects

The Vertical-Cavity Surface-Emitting Laser (VCSEL) is the ideal workhorse for optical interconnects. As low-cost, energy-efficient high-speed device which can be easily manufactured in large two-dimensional arrays and suits the power needs of common detector technologies directly modulated VCSELs can accomodate the bandwidth needs of optical interconnects for many years to come. Devices with modulation bandwidths beyond 30 GHz will be presented. To take benefit of advanced coding schemes for 100 Gbps per-channel data-rate, high-contrast gratings are needed to open up another degree of freedom in VCSEL device design. However, the numerical design is nontrivial and manufacturing has to be accurate.

Max Lemme, University of Siegen (max.lemme@uni-siegen.de)

Graphene and 2D Materials for Optoelectronics

Broad spectral range optical detection is of interest for technological applications such as imaging, sensing, communication and spectroscopy. Two-dimensional (2D) materials are very promising for such applications. Graphene is a suitable material for broadband detection because its linear dispersion relation allows carrier absorption across the entire spectrum from ultraviolet to terahertz. In this talk, graphene / silicon Schottky barrier diodes made of chemical vapor deposited (CVD) graphene on n-type Si substrates will be discussed. The effects of incident light intensity and wavelength are investigated. The diodes exhibit good rectifying behavior and high sensitivity to changes of incident light. A broad spectral response (SR) of 60 - 407 mA/W at a reverse dc bias of 2V is measured from ultraviolet (UV) to near infrared (NIR) light.

In contrast to graphene, molybdenum disulfide ($MoS2$) is an n-type semiconducting 2D material. Monolayer $MoS2$ has a direct band gap of ~1.8 eV, whereas bulk $MoS2$ has an additional indirect band gap of ~1.3 eV. In $MoS2$/Si diodes made with multilayer, CVD grown $MoS2$, the maximum observed spectral response is 8.6 mA/W. The spectral photoresponse measurement furthermore aids the understanding of the electronic structure of CVD grown $MoS2$ films, as the results show the fingerprint of the $MoS2$ band structure.

The comparison of the two diodes clearly demonstrates that spectral response measurements are an excellent tool to probe the electronic properties of novel 2-dimensional materials.

Jan Dubowski, Université de Sherbrooke (Jan.J.Dubowski@USherbrooke.ca)

Photocorrosion of GaAs/AlGaAs Heterostructures

Photocorrosion of semiconductors is typically considered a parasitic effect detrimental to functioning of electronic devices, such as field effect transistors, memory chips, light emitting diodes or electrochemical solar cells. The photocorrosion process depends on the concentration of impurities, energy and intensity of photons irradiating semiconductor microstructures, and on the nature of surrounding environment.

A typical defence against photocorrosion could include surface passivation with atoms of a material capable to neutralize chemical reactivity of the semiconductor surface, coating with hole scavenging compounds, or with oxide/nitride films to create a physical barrier between the semiconductor surface and a corrosion inducing environment. Under constant irradiation conditions and a continuously refreshing environment, photocorrosion of semiconductors is a well reproducible effect that could be monitored in situ with, e.g., photoluminescence (PL) or Raman spectroscopies.

I will review the results of our photocorrosion research of GaAs/AlGaAs microstructures immersed in different solutions, including deionized water and phosphate buffered saline and NH4OH solutions. By employing the PL effect, we were able to monitor photocorrosion with an in-depth nanometer-scale precision. We demonstrate that the super-slow photocorrosion processes can be successfully implemented for monitoring surface localized reactions involving bacteria, viruses and other electrically charged molecules.

Raphael Clerc, Institut d'Optique Graduate School (raphael.clerc@institutoptique.fr)

Solution Processable Organic Photodiodes

Solution processed organic photodiodes are one of the most promising application of organic electronics. After an introduction of the field, some aspects of the physics and modeling of the performance of organic photodiodes will be presented. In particular, some recent experimental and modeling results will be discussed. The first one is a detailed analysis of the impact of blend morphology of P3HT / PCBM devices, showing the critical impact of traps on performances, both on dark and illumination conditions. The second study investigates the impact of the "so called" oxygen doping. The impact of oxygen contamination will be discussed, and a simple experiment to detect it will be proposed.

Session O4: Silicon Photonics

Lawrence Chen, McGill University (lawrence.chen@mcgill.ca)

Silicon Photonics: Enabling Microwave Photonic Systems and Applications

Techniques for generating arbitrary RF waveforms are important for a broad range of applications. For example, UWB-over-fiber has recently emerged as a viable approach to extend reach and the area covered in high-speed wireless communications and sensing. As a second example, chirped microwave waveforms (and their subsequent compression) are useful in radar systems for increasing detection distance or resolution. Photonic techniques for generating UWB and chirped microwave waveforms provide increased flexibility, particularly with regards to reconfigurabilty and tunability. Additional application-specific advantages include compatibility with fiber optic transmission (UWB-over-fiber) or to achieve carrier frequencies of tens to hundreds of GHz with significant chirp rates (radar imaging).

In this presentation, we describe the use of silicon photonics for generating UWB and chirped microwave waveforms, thereby providing an integrated platform to enable a number of microwave systems and applications.

Jianping Yao, University of Ottawa (jpyao@uottawa.ca)

Silicon Photonics for Microwave Signal Processing

Silicon photonics is playing an increasingly important role in the implementation of photonic circuits for the generation and processing of optical and microwave signals. In this talk, techniques to generate and process optical and microwave signals using silicon photonic circuits will be discussed. Specifically, the use of a silicon-based on-chip optical spectral shaper to generate a largely chirped microwave waveform will be discussed. The use of an integrated sidewall phase-shifted Bragg grating to perform all-optical differentiation of an optical pulse, and the use of a silicon-photonic symmetric Mach–Zehnder interferometer incorporating cascaded microring resonators to achieve independently tunable multichannel fractional-order temporal differentiator will also be discussed.

Maurizio Burla, INRS-EMT (maurizio.burla@emt.inrs.ca; maurizio.burla@gmail.com)

with J. Azaña

Integrated Waveguide Bragg Gratings for Microwave Photonic Signal Processing: Current Trends and Future Prospects

In recent years, microwave photonics (MWP) signal processing using photonic integrated circuits has attracted a great deal of attention as an enabling technology for a number of functions not attainable by a purely electronic approach. In this context, waveguide Bragg grating (WBG) devices on silicon constitute a particularly attractive solution, thanks to their high compactness and flexibility in producing arbitrarily defined amplitude and phase responses, by directly acting on perturbations of the grating profile.

This talk will give a brief overview of recent advances in the field of integrated WBGs applied to microwave photonics, with emphasis on a number of recent demonstrations by our group of the use of WBGs for microwave and ultrafast optical signal processing. Examples of applications include wideband tunable RF filters, broadband true-time-delay lines and phase shifters, THz-bandwidth pulse shapers, and instantaneous frequency measurement systems on-chip. Finally, a perspective on the exciting possibilities offered by the silicon photonics platform in the field of MWP will be given, potentially enabling integration of highly-complex active and passive functionalities with high yield on a single chip, with a particular focus on the use of WBGs as basic building blocks.

Xiaoguang (Leo) Liu, University of California, Davis (lxgliu@ucdavis.edu)

with H. Rashtian, B. Yu and J. Gu

High-Speed CMOS-based THz Interconnect with Micromachined Silicon Dielectric Waveguide

Although the absolute speed of current general purpose microprocessors has slowed down in progression due to their increasing power consumption and thermal dissipation bottleneck, today's computing systems, be it a personal computer, a supercomputer, or a large data-center, has roughly maintained performance scaling by taking advantage of parallelism. While alleviating the requirement on each individual processing unit, parallelism shifts the burden to the interconnects between processors and cores. The ever-increasing inter- and intra-chip communication bandwidth imposes a big challenge over decades: interconnect bottleneck.

Both electrical interconnects, in the form of metal traces and bondwires, and optical interconnects, in the form of integrated photonic transceivers, have been extensively studied over the last few decades. Electrical interconnects can be easily integrated within current semiconductor design and fabrication processes but face increasing challenges in terms of the usable bandwidth and signal integrity issues. Optical interconnects offer extremely large data bandwidth and very low distortion but suffer significant difficulties in integration with existing semiconductor technologies.

THz interconnect (TI) aims to bridge the gap to enable energy efficient centimeter range communications, and complement EI and OI by integrating both electronic and optic advantages. In this work, we demonstrate the feasibility of using the terahertz frequency band for ultra-high speed chip-to-chip interconnects. In particular, we will demonstrate the use of micromachining technologies for the fabrication of low-loss large-bandwidth single- and multi-mode THz interconnects that can be realistically integrated with existing semiconductor fabrication and packaging processes. We will also demonstrate critical CMOS-based circuit components for THz interconnects.

David Plant, McGill University (david.plant@mcgill.ca)

Silicon Photonics Enabled 400G/1T Short Reach Optical Interconnects

Silicon Photonics (SiP) is a rapidly evolving technology that is fabricated using CMOS manufacturing processes and enables device and systems designers to build dense photonic circuits on silicon. Silicon modulators operating near 1300 nm are desirable for telecommunication systems for close-to-zero dispersion in short distance high-speed transmission. Higher order intensity modulation (IM) with direct detection (DD) on a single wavelength is an attractive approach to efficiently increase the bit rate throughput without the expensive capital expenditure required by multi-wavelength multiplexing. Pulse amplitude modulation (PAM) has been shown to be a promising candidate for intensity modulation in optical short reach communications. The performance and throughput of direct detection can be improved by employing an analog-to-digital converter (ADC) after the PIN+TIA receiver with filtering from feed forward equalization (FFE).

In this presentation we study the transmission performance of PAM-4 and PAM-8 formats at 112 Gb/s using an 8-bit DAC and show how DACs can greatly benefit PAM formats by employing more output levels to generate the PAM waveform. We also discuss 224 Gb/s transmission over 10 km using a single 1310 nm laser. To achieve this bit rate on a single wavelength, IM with a 56 Gbaud PAM-4 signal in conjunction with PDM is employed, where polarization demultiplexing at the receiver is performed using a novel DSP that operates in the Stokes space after a DD receiver. When the proposed PDM IM/DD system is used in conjunction with WDM, it requires using only two and five laser sources to achieve aggregate data rates of 400 and 1 Tb/s, respectively; half of the required number of lasers in an equivalent single polarization system.

COFFEE BREAK (GEORGIA FOYER)

****

Trevor James Hall, University of Ottawa (thall@site.uOttawa.ca)

Photonic Integrated Circuits for Electro-optic Microwave Frequency Multiplication and Frequency Translation: Spurious Harmonics Suppression by Design

In the past two decades or so there has been a plethora of publications in the field of microwave photonics that have described essentially the same generalized Mach-Zehnder Interferometer (MZI) circuit architecture: a 1×N splitter directly interconnected to a N×1 combiner via an array of N electro-optic LiNbO3-based phase modulators, each adapted to particular design goals. The applications have generally been to single-side-band (SSB) modulation or electro-optic microwave signal frequency multiplication. The difference between the circuits proposed have largely concerned variations of the static optical and electrical phase shifts required or the implementation of an equivalent circuit using standard Mach-Zehner modulators (MZM) rather than individual phase-modulators (PM) as the basic building brick.

In this paper, I present a methodology that that specifies the architecture required to meet specified design objectives such as the suppression of unwanted products. Moreover, it shown how to use the intrinsic phase relations between the ports of splitters and combiners and specifically multi-mode interference (MMI) couplers to implement the static optical phase shifts required by these circuits, thereby avoiding the need to apply static DC bias to the electro-optic modulators and the associated drift issues that otherwise require complex stabilization circuitry. Circuits capable of single-side-band suppressed-carrier modulation and frequency octupling show a simulation performance equal to or better than results reported in the literature. In particular, a new cascade architecture implementation is reported that offers a 50% lower optical insertion loss and a 50% reduced RF power drive requirement compared to previously known circuits. While LiNbO3 technology offers a mature solution to the small scale integration of MZM structures, this work anticipates photonic integrated circuits based on Si and/or InP material integration platforms emerging as the preferred choice. In this context the continuous advances made in improved speed, linearity, footprint, and energy consumption of electro-optic phase modulator devices in both material platforms augurs well.

Sébastien Rumley, Columbia University (sr3061@columbia.edu)

Understanding Design Trade-offs of Ring Resonator Based Silicon Photonics Links

Optical ring resonators can be utilized to perform modulation, switching and filtering in photonics links. Microring based modulators have been shown to achieve high bit-rate with high energy efficiency. Although optimizing single device performance metrics is an important research area, another research goal is to design parameters to optimize link-wide metrics as total link bandwidth or energy-per-bit. In this talk, we present several models permitting to obtain such link-wide vision. We then show how these models can be exploited to identify equilibrated link designs. We finally use these results to conclude on the absolute capabilities of silicon photonics links.

Bahram Jalali, University of California, Los Angeles (jalali@ucla.edu)

Optical Information Capacity of Silicon

Modern computing and data storage systems increasingly rely on parallel architectures where processing and storage load is distributed within a cluster of nodes. The necessity for high-bandwidth data links has made optical communication a critical constituent of modern information systems and silicon the leading platform for creating the necessary optical components. While silicon is arguably the most extensively studied material in history, one of its most important attributes, its capacity to carry optical information, has only recently been investigated. The calculation of the information capacity of silicon is frustrated by nonlinear losses, phenomena that emerge in optical nanowires as a result of the concentration of optical power in a small geometry. Nonlinear losses are unique to silicon and are absent in silica glass optical fiber and other common communication channels. They lead to new types of noise and fluctuations that limit the information capacity well before the loss itself becomes appreciable.

This talk will present the information capacity of silicon, explains its origins, and outline solutions for extending it. The amount of information that can be transmitted by light through silicon will be a key consideration in future information systems and will impact the roadmap of silicon photonics. The talk will also highlight new technological trends and emerging applications in the field of photonics including real-time optical data compression and analytics.

Sébastien Le Beux, École Centrale de Lyon (sebastien.le-beux@ec-lyon.fr)

Thermal Aware Design Method for VCSEL-based On-Chip Optical Interconnect

Optical Network-on-Chip (ONoC) is considered as one of the key solutions for future generation on-chip interconnects. However, silicon photonic devices in ONoC are highly sensitive to temperature variation, which leads to a lower efficiency of Vertical-Cavity Surface-Emitting Lasers (VCSELs), a resonant wavelength shift of Microring Resonators (MR), and results in a lower Signal to Noise Ratio (SNR). In this talk, a methodology enabling thermal-aware design for optical interconnects relying on CMOS-compatible VCSEL is presented. Thermal simulations allow designing ONoC interfaces with low gradient temperature and analytical models allow evaluating the SNR.

Roel Baets, Ghent University (roel.baets@intec.ugent.be)

Silicon Photonics for Applications in Life Science

Silicon photonics takes advantage of the maturity of CMOS technology to implement complex photonic functionality and is best known for its application in optical interconnect, datacom and telecom, where it is rapidly moving into industrial deployment. But this technology may also be game-changing for many other photonics applications with a reasonable volume or a cost-critical market. This is particularly true for applications in life science. These include lab-on-chip applications where the chip is a consumable and also in point-of-care medical applications where affordability and portability of an instrument is very important. A broad range of such applications will be discussed.

Emmanuel Hadji, CEA-INAC (emmanuel.hadji@cea.fr)

with C. Renaut, B. Cluzel, E. Picard, J. Dellinger, D. Peyrade and F. de Fornel

Silicon Based Opto-fluidics: On-chip Optical Tweezing

We report here on-chip optical trapping and handling of micrometer-sized dielectric particles by SOI nanobeam cavities. We achieve single nanoparticle trapping as well as multiple particles assembling. Using a new architecture of near-field coupled nanobeam cavities, we also evidence particle motion control and handling between specific trapping sites. These results are based on a new an unprecedented opportunity to mold on a chip the morphology of the optical field by wavelength-induced optical-field tuning. They open exciting perspectives for future development of nanoscale optofluidics with configurable optical traps, sensors or dynamic nano-tweezers.

Track S: Bioelectronics, Radiation & Sensors

Session S1: Bioelectronics

Adam Woolley, Brigham Young University (awoolley@chem.byu.edu)

with B. Uprety, E. P. Gates, K.V. Lee, J.K. Jensen, S. Noyce, N.B. Prestwich, R.C. Davis and J.N. Harb

Exploiting the Interface Between Biology and Electronics

Living systems spontaneously create amazingly complex structures with nanoscale dimensions, such as proteins, nucleic acids, etc. If a fraction of the sophisticated capabilities of Biology could be utilized for the making of inorganic structures for electronics applications, it might well be possible to take a quantum leap forward, far beyond the limits of current top-down methods. We are constructing scaffolded DNA origami nanostructures and selectively depositing inorganic materials onto these DNA templates. We have characterized these bottom-up-fabricated systems and studied their electrical properties. These conductive DNA assemblies have excellent potential to be utilized in nanofabrication, electronics, sensing, photonics, etc.

Hassan Ghasemzadeh, Washington State University (hassan@eecs.wsu.edu)

Sustainable Smart Health: Technology Self-Management to Enhance User Compliance

Recent years have witnessed considerable research demonstrating the potential of wearable sensors in healthcare and wellness applications. Current wearable devices, however, are difficult to use by the elderly, and produce unreliable and inconsistent measurements when deployed in uncontrolled environments. This talk presents two techniques, self-configuration and reliability, in networked wearables. The self-configuration component refers to the autonomous learning of computational algorithms in response to changes in user's environment. The reliability component attempts to enhance robustness of the system to mitigate for the physical unreliabilities, while taking into account stringent constraint resources available on the wearable sensors.

Bozhi Tian, University of Chicago (btian@uchicago.edu)

Designing Complexity in Silicon Nanostructures for Enhanced Bio-interfaces

Biological systems are rich with electrical and mechanical activities. Silicon based nanoscale materials and devices represent diverse and powerful tools for achieving nanoscale bioelectric interfaces with cells and tissue. My talk will focus on several biomimetic design considerations toward breaking down the boundary between nonliving and living systems across multiple length scales. I will describe how we experimentally apply these designs in the nanoelectronic systems for building minimally invasive interfaces with single cells and subcellular components. Specifically, I will discuss a new synthetic approach to realize the biomineral-like probes for extracellular measurements.

Ling Qin, Harvard University (qinl06@gmail.com)

with J. Abbott

A Solid-state Platform for Neurotechnology

One long-standing challenge in experimental neuroscience is the lack of electrophysiological tools that can perform both intracellular and parallel measurements of electrical activities of a mammalian neuronal network. Both Intracellular and parallel measurements are highly desired, because the former enables high-fidelity signal recording/stimulation while the latter is necessary for simultaneous monitoring of a large number of neurons as a network. Here we present our ongoing research to develop a nano-bio interface array on a solid-state platform, which is aimed at achieving such dual capabilities. This work is a collaboration between Hongkun Park's group and Donhee Ham's group at Harvard University.

Boris Stoeber, University of British Columbia (boris.stoeber@gmail.com)

with I. Mansoor, S. Ranamukhaarachchi and U. Hafeli

Metal Microneedles for Biomedical Applications

Microneedles are biomedical microdevices that provide a pathway across the skin barrier to exchange fluids or compounds with the body for drug delivery or biosensing. The effectiveness of microneedles for transdermal drug delivery has been demonstrated in clinical trials, and it has been shown that this drug delivery method can be painless. We have developed an inexpensive fabrication processes based on solvent casting and electroplating to fabricate arrays of hollow microneedles from metal. Solvent casting leads to polymer microstructures on a mold surface as the solvent evaporates. This process involves interesting flow physics inside the sub-millimetre thick fluid film.

Olukayode Karunwi, Clemson University (okarunw@clemson.edu)

with F. Alam and A. Guiseppe-Elie

Biofabrication, in vitro and in vivo Performance of Dual Responsive Glucose and Lactate Implantable Biosensor in a Piglet Trauma Model

Implantable biosensors for the continual measurement of interstitial glucose and lactate are being developed. Fabrication of oxidoreductase enzyme-rich biorecognition membranes deposited via pyrrole electropolymerization at microfabricated electrodes has been achieved. A catalytic layer of Ni-hexacyanoferrate placed at the electrode-enzyme interface for enhanced peroxide response produced a 20-fold increase (14.19 nA vs. 0.7 nA) in buffered H_2O_2 measured at 650 mV vs. Ag/AgCl. In vitro characterization showed a sensitivity of 0.68 mA/cm2/mM and 0.36 mA/cm2/mM and a limit of detection of 0.05 mM and 7.9 mM for glucose and lactate respectively.

Session S2: Sensors

Sameer Sonkusale, Tufts University (sameer@ece.tufts.edu)

CMOS Imagers for Lifetime and Multispectral Measurements

Michael Suster, Case Western Reserve University (mas20@case.edu)

A Miniaturized Platform for MHz-to-GHz Dielectric Spectroscopy

John Madden, University of British Columbia (jmadden@ece.ubc.ca)

with M. Saquib Sarwar, Y. Dobashi, E. Glitz, E. Cretu and S. Mirabbasi

Piezolonic Sensors for Transparent Touch Sensitive Displays

Application of pressure to materials containing ions leads to the generation of voltages and currents, analogous to effects observed in piezoelectrics. A key difference is that the ions that are displaced are mobile in piezoionics, leading to very large charge transfer compared to piezoelectrics, at lower voltages. The piezoionic concept is introduced, and we demonstrate its application to a flexible, stretchable and transparent touch sensor.

Peter Grutter, McGill University (peter.grutter@mcgill.ca)

Measuring Electronic Properties on a Nanoscale using Atomic Force Microscopy

Electrostatic force microscopy, a variant of atomic force microscopy, can play an essential role in characterizing nanoelectronic devices. I will present a brief summary of our capacities developed over the past few years (see www.physics.mcgill.ca/~peter for details):
1. Characterization of the spatial and temporal fluctuations of the surface potential as well as charge traps on semiconductors and oxides.
2. Measurements of the potential "landscape" have been performed on functioning graphene and nanotubes devices.
3. Measurement of the electronic structure (including single electron charging and level degeneracies) of individual and coupled self-assembled quantum dots of InAs and Au.

Shankar Rananavare, Portland State University (ranavas@pdx.edu)

with H. Tran

Mechanism of Sensing of Reducing and Oxidizing Gases on 1D Nanostructures

In this talk, I will present our latest studies of molecular mechanisms of sensing in sensors based on nanowires and nanotubes. N-doped SnOx nanowires selectively detect chlorine which is a strong oxidizing gas. Migration of chlorine in the interstitial vacancy sites of the oxide results in slow recovery times for the sensor. Palladium nanoparticle decorated multiwall nanotubes are used to detect reducing gas, hydrogen. The latter system shows a negative differential resistance resulting from a carrier inversion phenomenon.

COFFEE BREAK (GEORGIA FOYER)

****

Nicola Guerrini, Science and Technology Facilities Council (nicola.guerrini@stfc.ac.uk)

CMOS Image Sensors for TEM: State of the Art Detectors and Outlook

CMOS image sensors have revolutionised TEM (Transmission Electron Microscopy). Integrating CMOS technology with TEM allows direct detection and fast image acquisition, whilst also delivering better sensitivity. The result of the radical change represented by CMOS Image Sensors is that researchers from all over the world are collecting and publishing new and exciting results every day. This paper will first present an overview of the current state of the art and will then provide insight into the challenges and opportunities that the future might hold for the TEM world.

Markus Beckers, RWTH Aachen University (Markus.Beckers@ita.rwth-aachen.de)

with C.A. Bunge, H. Poisel and T. Gries

Sensor Applications Based on Standard and Newly Developed Polymer Optical Fibres

The talk covers approaches for optical sensors based on polymer optical fibres. We will show sensors based on classic step-index POFs, but we will also demonstrate a newly developed fibre-fabrication process and novel applications based on this new fibre type.

Ikhwana Elfitri, Andalas University (ikhwana@ft.unand.ac.id)

Universal and Scalable Spatial Audio Coding

The first part of this talk discusses recent progresses in spatial audio coding while the second part presents a new approach which is universal for various applications and scalable for different operating bitrates. Interestingly, this method is also promising for interactive rendering or object-based audio reproduction.

Zane Bell, Oak Ridge National Laboratory (bellzw@ornl.gov)

Neutron Detection

The neutron was hypothesized in 1920 by Rutherford and "discovered" in 1932 by Chadwick, who deduced their presence from measurements of recoiling gas atoms and protons. Although this was done more than 80 years ago, the basic detection principles used by Chadwick remain in use today: Neutrons, being massive neutral particles interacting with matter via the strong nuclear force, are always detected by observing the directly or indirectly produced ionizing charged particles. In this presentation, common and some uncommon detection techniques will be reviewed. These include proton recoil, fission, gamma production from scattering and capture, activation foils, track etch detectors, photographic emulsions, cryogenics, Cherenkov light, and self-powered detectors. In addition, the properties making materials desirable or suitable for use in neutron detectors will be discussed.

Oak Ridge National Laboratory is managed by UT-Battelle, LLC under Contract No. DE-AC05-00OR22725 with the U.S. Department of Energy.

Fabrice Retiere, TRIUMF (fretiere@triumf.ca)

Scintillation Light Detection in Cryogenic Liquids with CMOS and Other Silicon-based Photo-detectors

Efficient detection of scintillation light in noble-gas liquid is a key requirement to enable the characterization of neutrinos and the identification of dark matter particles. Silicon-based photo-detectors are a compelling solution as opposed to photo-multipliers because of their high efficiency, low dark noise rate at low temperature, and low intrinsic radioactivity. So-called silicon photo-multipliers that rely on Geiger-mode avalanche are the primary candidate to equip next generation experiments. We will review the performances of SiPMs and discuss in particular the serious issue of capacitance per unit area.

Session S3: Biosensors

Mehmet Dokmeci, Massachusetts Institute of Technology (mehmetd@MIT.EDU)

Microscale Sensors and Systems for Tissue Engineering and Regenerative Medicine Applications

The advances in micro and nanoscale sensors are enabling niche applications. Recent proliferation of organ on a chip platforms has resulted in vitro living systems representative of the body conditions. Ultra sensitive and miniature physical (ph, temperature, oxygen) sensors and biochemical sensors are being developed to monitor the environment and activity of these organ constructs. The sensor chips can be integrated with the bioreactor substrates for continuous monitoring. The applications of these devices can be extended to other biomedical fields such as wound monitoring as well as in environmental monitoring such as portable detection platforms.

Howard Wikle, Auburn University (hcw0002@auburn.edu)
with B. Chin

Microfabricated Magnetoelastic Biosensors

Magnetoelastic biosensors are a type of free-standing, mass-sensitive, acoustic wave sensors used to detect pathogens responsible for human diseases. These biosensors are based on magnetoelastic resonator platforms that are microfabricated to achieve uniform material properties and structural dimensions down to tens of micrometers. These platforms are coated with a biorecognition element, highly specific bacteriophage in this instance, to bind and capture bacteria including Salmonella or Staphylococcus. Both the operation frequency and mass sensitivity increase as the dimensions shrink, allowing detection of these and other chemicals or pathogens at concentrations that are dangerous to human health.

Luca Selmi, Università degli Studi di Udine (luca.selmi@uniud.it)

Modeling and Simulation of CMOS Nanoelectronic Biosensors

Integrated electronic biosensors are attracting increasing interest as enablers of cheap, pervasive, label-free, massively parallel and ultrasensitive detection for e-health, personalized medicine, safety and security applications. Among various biosensor concepts, on chip impedance spectroscopy is especially promising, but further developments in the field demand a detailed understanding of the signal transduction chain and careful interpretation of experimental results. Modeling and simulation is instrumental to achieve all this and enable optimized chip designs for reliable operation. The talk addresses recent developments of modeling and simulation for integrated impedimetric biosensors, and the lessons learned from comparison with well controlled experiments.

Kalle Levon, New York University (kalle.levon@gmail.com)

Ion Sensitive Floating Gate FET Array with Polyaniline Coating

Electrically conducting polymers are suitable for surface bioconjugation strategies due to the convenient organic chemistry routes but also as advantageous hybrid materials for microelectronic biosensors due to their high bandgap sensitivity, and possibilities for nanoscale surface area formation. We shall present how organic conductors can be used to functionalize ion sensitive floating gate field effect transistors (ISFGFETs) designed to measure biological binding events. Our conductive polymer modified ISFGFET sensor arrays are a promising alternative to potentiometric biosensors due to their signal amplification, high throughput and scalability advantages.

Hua Wang, Georgia Tech (hua.wang@ece.gatech.edu)
with J.S. Park and T. Chi

A Multi-Modality CMOS Sensor Array for Cell-Based Assay and Drug Screening

We present the world-first 4-modality pixelated CMOS sensor array for cellular assay and drug screening with no special post processing. Implemented in a standard 130nm CMOS process, the array has 144 tri-modality pixels (voltage, impedance, optical shadow sensing) with circuit sharing and 9 temperature sensors for joint-modality holistic cellular characterizations. On-chip mouse neurons show concurrent optical and impedance multi-modality imaging for cell localization and cell-CMOS surface attachment monitoring. Human cardiomyocytes demonstrate cardiac drug detection by joint-modality cellular potential and impedance detection.

Session S4: Radiation Effects and Detection

Kate Shanks, Cornell University (ksg52@cornell.edu)

X-ray Area Detectors for High Dynamic Range Imaging

Several classes of experiments at accelerator-based x-ray light sources, ranging from imaging to diffraction, require area detectors with high dynamic range, i.e. low read noise combined with a large well depth. Using hybrid detector technology in which a thick, high-resistivity semiconductor sensor is bonded to a custom readout ASIC fabricated in commercial CMOS, we combine analog and digital processing to achieve a high dynamic range, while improving on the count rate limitations imposed by a purely photon counting approach. Pixel design will be described and examples from experiments will be shown to demonstrate detector capabilities.

Klaus-Peter Ziock, Oak Ridge National Laboratory (ziockk@ornl.gov)

Data Fusion of Visual, LWIR, and Gamma-Ray Imaging for Nuclear Security

Gamma-ray detection is universally applied for the detection and control of illicit nuclear materials. More recently, gamma-ray imaging has been explored as a means to locate and determine the distribution of such materials. Overlays of gamma-ray and visible-light images are the simplest forms of data fusion, providing spatial cues to aid efforts in nuclear security. We have been applying advanced data fusion using detector location, and visual and long wave infrared (LWIR) images to inform collection of static gamma-ray images of moving targets; allowing one to overcome limiting systematic background fluctuations while uniquely localizing radiation signatures to passing targets.

Antoine Touboul, University of Montpellier (antoine.touboul@ies.univ-montp2.fr)

Radiation Effects in Power Applications

Power devices are the core of energy conversion systems and are used in many sectors (consumer electronic, avionic, automotive). Although radiation effects are known to mainly affect integrated electronics, the high electric field inside power devices make them sensitive as well to radiation. In the frame of the smart vehicle development, it is now mandatory to determine how the natural radiative environment can impact the reliability of automotive and to propose new concept for the selection and the qualification of new generation devices dedicated to automotive.

Mokhtar Chmeissani, Institut de Física d'Altes Energies (mokhtar@ifae.es)

3D Semiconductor Sensor For Position Emission Tomography

Positron Emission Tomography (PET) is nuclear medicine diagnostic tool for molecular imaging. Though the spatial resolution of the current state of the art PET system is few millimeters, the PET possesses excellent sensitivity to detect the molecular functionality and this is what makes PET device unique specially for detecting cancer. However the sensitivity of the current state of the art PET is limited by:
1- scattered photons;
2- large volume crystal;
3- detection efficiency for the 511keV photons.

The Voxel-Imaging-PET project (VIP) is developing a new sensor based on High-Z pixel semiconductor detector (CdTe) bonded to thinned frontend electronics and stacked on the top of each other, yielding a true 3D module that can detect the 511keV at any depth while maintaining the same energy and spatial resolutions. Such compact geometry yields more than 400 voxel/cm3, with each Voxel having its complete readout electronics (preamp, trigger, shaper, peak-hold, ADC, and TDC). GAMOS simulation shows that with the VIP approach the spatial resolution can reach the intrinsic limits set by the physics of PET.

Frédéric Wrobel, University of Montpellier (frederic.wrobel@ies.univ-montp2.fr)

Simulation Tools for Soft Error Rate Calculations in SRAM

Alpha particles emitted by natural radioactive impurities in integrated circuits are among the main sources responsible for soft errors in CMOS technologies. In order to investigate the role of these impurities on the error rate, we present the mechanisms at play and the way we can simulate these disintegrations within the electronic device. The main contribution is due to the decay chains of Uranium and Thorium that have respectively 12 and 10 radionuclides. The knowledge of the state of each of these decay chains is crucial for predicting the Soft Error Rate.

COFFEE BREAK (GEORGIA FOYER)

****

Minoru Fujishima, Hiroshima University (fuji@hiroshima-u.ac.jp)

Evaluation and Modeling of Terahertz CMOS Devices

Terahertz (THz) CMOS devices will be a key to personalize ultrahigh-speed communication and non-invasive sensors. In order to utilize the THz CMOS devices, we have to start from building device models for circuit simulation. However, when operation frequency becomes over 100-GHz range, which is THz, we have to consider physical backgrounds which have not been well discussed by CMOS engineers. For example, what is a good reference for measurement? How should characteristics of a target device be extracted precisely from measured results? How do we build the device model which can express precisely nonlinear response? In this presentation, measurement, evaluation and modeling are discussed for utilizing CMOS devices in circuit design.

Salvador Pinillos Gimenez, Centro Universitário da FEI (sgimenez@fei.edu.br)

An Innovative Non-Standard Layout Style (Diamond) to Boost the Electrical Performance and the Radiation Tolerance of MOSFETs, Focusing on Space and Medical CMOS ICs Applications

This paper aims to describe an innovative layout style, named Diamond (hexagonal gate geometry), which it is capable to boost the electrical performance and, in the same time, the Total Ionizing Dose (TID) tolerance of Metal-Oxide-Semiconductor Field Effect Transistors (MOSFET), without burdening the current planar Complementary MOS (CMOS) Integrated Circuits (ICs) manufacturing process. This manuscript also presents the main layout features of the Diamond MOSFET, the new effects observed in this innovative transistor structure and the modeling proposed to describe its electrical behavior. Some experimental results are illustrated to evidence its potential use in space and medical CMOS ICs applications.

Maurice Garcia-Sciveres, Lawrence Berkeley National Laboratory (mgarcia-sciveres@lbl.gov)

Data Compression Efficiency in Silicon Detector Readout

This talk will present recent work applying information theory concepts to analyze the readout of silicon strip and pixel detectors commonly used in particle physics. While efficient data extraction from scientific detectors ha always been a concern, the transition to purely digital data flow is relatively recent, where analog signals only exist within a few mm of the sensor elements they are produced in. Efficient extraction of digitized data is now a major concern for planned future detectors, which will be larger (many square meters) and at the same time have higher data rates per unit area than present examples. Achievable output bandwidth is one of the limitations on the performance of future detectors. At the same time, thanks to Moore's Law, substantial processing power that can be incorporated in the circuits generating digital signals. An understanding of how data compression methods apply to the specific case of silicon detector data is therefore important.

Amin Arbabian, Stanford University (arbabian.a@gmail.com)

RF and Microwave Imaging with Applications in Medicine

Conventional medical imaging modalities (MRI, PET, CT) rely on expensive and bulky hardware that limit usage to hospitals and clinics. Large and expensive devices prohibit usage in ambulatory cases or where time-sensitive and on-site diagnosis is critical (e.g., detection of internal hematomas or intracranial abnormalities). In addition to access, safety is a concern with a majority of the imaging techniques. This is especially an issue in applications like cancer screening where frequency of test is limited by concerns with radiation dosage.

This presentation will provide an overview of several RF/Microwave imaging techniques for detection and reconstruction of dielectric contrast created by hemorrhages or by tumor angiogenesis and necrotic cores. Imaging results from various dielectric materials will be outlined. Additionally, a new non-contact imaging technology that can be used for remote interrogation of hidden/embedded objects in highly dispersive media will also be presented.

Jean-Luc Autran, Aix-Marseille University (jean-luc.autran@univ-amu.fr)
with P. Roche, G. Gasiot and D. Munteanu

Natural Radiation in Bulk CMOS and FD SOI

Natural radiation is today a major integrated circuit reliability issue, as important as intrinsic failure modes, and especially for high reliability applications (medical, automotive, transportation). Current deca-nanometer CMOS technologies are characterized by an increasing sensitivity to background radiation, including atmospheric particles (neutrons, protons, muons) and alpha particles emitted from traces of contaminants in circuit and packaging materials. This presentation will survey our recent real-time experiments, conducted in altitude and underground to characterize soft error mechanisms in memory circuits. Recent advances concerning the modeling and simulation of these soft errors will be also presented, as well as perspectives for FD-SOI technologies.

Anton Tremsin, University of California, Berkeley (ast@ssl.berkeley.edu)
with J.V. Vallerga, J.B. McPhate, O.H.W. Siegmund and R. Raffanti

Novel Electronics for the Neutron Imaging Detectors and Their Applications

The latest developments of the fast parallel electronics developed for the quad Timepix readout enable a number of unique non-destructive testing techniques. This readout combined with Microchannel Plates (MCPs) provides the possibility to register both position (with resolution of 15-55 μm) and time (with resolution of ~20 ns) of multiple nearly simultaneous events. The MCPs convert a single photon/ electron/ ion/ neutron into a charge of 10^4-10^7 electrons localized within 4-10 μm from the event position. The application of this detection technology at the neutron beamline facilities enables unique studies in materials sciences, energy research, geological sciences, cultural heritage, biomedical research and many other fields.

In this paper we present a detector with 28x28 mm2 active area. The fast parallel readout electronics enabling readout speeds of up to ~1200 frames/sec is a crucial component of the detection system. The MCP-Timepix detector can be very attractive for applications where high spatial resolution needs to be preserved for nearly simultaneous events, e.g. time of flight measurements with pulsed sources. The low noise of the Timepix readout enables operation at gains as low as 10^4-10^5, which should extend the lifetime of the detectors operating at high counting rates.

Session S5: Bioelectronics

George Malliaras, École des Mines de Saint-Étienne (malliaras@emse.fr)

Interfacing with the Brain Using Organic Electronics

One of the most important scientific and technological frontiers of our time lies in the interface between electronics and the human brain. Interfacing the most advanced human engineering endeavor with nature's most refined creation promises to help elucidate aspects of the brain's working mechanism. It also promises to deliver new tools for diagnosis and treatment of a host of pathologies including epilepsy and Parkinson's disease, as well as new capabilities through brain/machine interfaces. Current solutions, however, are limited by the materials that are brought in contact with the tissue and transduce signals across the biotic/abiotic interface. The field of organic electronics has made available materials with a unique combination of attractive properties, including mechanical flexibility, mixed ionic/electronic conduction, enhanced biocompatibility, and capability for drug delivery. I will present examples of organic-based devices for recording and stimulation of brain activity, highlighting the connection between materials properties and device performance. I will show that organic electronic materials provide unparalleled opportunities to design devices that improve our understanding of brain physiology and pathology, and can be used to deliver new therapies.

Jennifer Gelinas, New York University, Langone Medical Center (jennifer.gelinas@nyumc.org)

Large-scale Neural Interface Devices

Recording from neural networks at the resolution of action potentials is critical for understanding how information is processed in the brain. We developed an organic, ultra-conformable, biocompatible and scalable neural interface array (the "NeuroGrid") that can record both LFP and extracellular action potentials from cortical neurons without penetrating the brain surface. Spiking activity demonstrated consistent phase modulation by ongoing oscillations and was stable in recordings exceeding one week. We also recorded LFP-modulated spiking activity intra-operatively in patients undergoing epilepsy surgery. The NeuroGrid constitutes an effective method for large-scale, stable recording of neuronal spikes in concert with local population synaptic activity.

Jacob Robinson, Rice University (robinson.jacob@gmail.com)

Nanotechnology for Reading and Writing Neural Activity

Nanofabricated electronic, magnetic, and photonic devices can help us understand computation in the brain by precisely reading and writing activity in individual neurons. In this talk I will describe some of our latest work to develop next-generation neural interfaces using technologies borrowed from semiconductor nanofabrication.

Sepideh Mazrouee, University of California, Los Angeles (sepideh@cs.ucla.edu)

Optimization Techniques for DNA Phasing, Above and Beyond

The term Haplotype refers to a set of DNA variations that tend to be inherited together in a single chromosome as a unit. Haplotype assembly problems aim to reconstruct distinct copies of an organism's chromosome. Organisms in general can be haploid, diploid or polyploid. While reconstructing chromosome copy for a haploid organism might be senseless, it has various applications for diploid and polyploids. This reconstruction problem has multiple methods as well as many computational and experimental difficulties and has a wide range of application from understanding the evolution history in creation of different species to important applications for human in prognosis of inheritable diseases, personalized drug design and even transplant matching in genetic level. Understanding structure of an individual's haplotype plays a significant role in various fields of genetics. Next generation sequencing technologies have enabled us to reconstruct an individual's haplotype copies using DNA short fragments. The quality of the assembly, however, is affected by the quality of the sequenced data, making accurate and fast reconstruction of the haplotypes highly challenging. The objective, therefore, is to design algorithms that mitigate this impact and rebuild the most likely copies of each chromosome accurately and fast. This necessitates development of algorithms that not only reconstruct haplotypes accurately but also require low computation time and therefore enable scalability of the reconstruction process.

Dorothée C.M.C. Petit, University of Cambridge (dcmcp2@cam.ac.uk)

Interface of Nanomagnetism and Cancer Therapy

In recent years there have been several suggestions and in-vitro demonstrations of how nanomagnets can be used to treat cancer tumours. Most schemes rely on hyperthermia in which the magnetic hysteresis of the nanomagnets is used to apply localised heating. A different approach uses the torque exerted on an anisotropic nanomagnet to apply destructive forces to cell walls, thus triggering cell death. In this talk we present for the first time an in-vivo demonstration of cancer cell killing using mechanical torques generated by lithographically defined nanomagnetic structures with carefully designed magnetic properties. The structures are prepared on a silicon wafer using standard lithography and thin film deposition and are then lifted off into solution and injected into glioblastoma (an aggressive brain cancer tumour) in mice. The mice are exposed to a rotating magnetic field daily and the progress of the tumour and health of the mice monitored over several weeks. The cohort of mice treated with nanomagnetic structures and applied field shows significantly increased survival times, increased body weight and reduced tumour size compared to control groups, thus indicating that nanomagnets are a potentially important and potent new cancer therapy.

Alistair McEwan, University of Sydney (alistair.mcewan@gmail.com)

Electrical Impedance Spectroscopy – A Biomarker for Implant Systems

One of the major bottlenecks in advancing innovative neurotechnologies is selectivity in electrical stimulation and sensing. The engineering problem here is understanding the impact of the complex tissue on the transfer function of the system and its dynamic changes. While it has been established that the frequency dependent capacitance has a major effect, it is rarely included in models. It is still unknown how tissue dielectric properties change with stimulation, and new biomarkers are required.

The transfer function should be characterized over a physiologically relevant frequency range that spans the expected recorded signals and typical bi-phasic stimulation waveforms. The essential information needed is the dielectric properties of the tissue and the anatomy and anisotropy. This will allow impedance matching of the tissue load to the stimulus, optimising stimulation efficiency, current steering and waveform matching to optimise stimulation. Two innovative approaches here are:
1. developing improved FEM based models of bulk tissue dielectrics using values obtained by new methods such as bioimpedance tensor spectroscopy, fast neural electrical impedance tomography and magnetic resonance electrical impedance tomography; and
2. Measuring the electrical impedance spectrum transfer function seen by implanted electrode systems by taking advantage of supporting hardware that is starting to appear such as the Medtronic PC+S.

Advances in these areas may lead to improved electrode placements, current steering strategies, optimal stimulation waveforms, improvements in inverse source detection and fundamental understanding of neural systems.

Session S6: Wireless

Baptiste Grave, Stanford - IEMN ISEN Lille (bgrave@stanford.edu)

Subsampling Techniques applied to Gbps Wireless Communications

This talk shows how the properties of subsampling can be used in Gbps wireless communications. Downconversion, IQ demodulation and out of band filtering can be merged in a single operation, resulting in a simple and optimized RF sub-sampler for mmW communications. As a proof of concept, an IF (20 GHz) to DC receiver for 802.11ad (WiGig – 1.75 GSps) was designed in 28 nm CMOS. It shows that subsampling is compatible with wide band RF receivers in modern CMOS technologies.

Julian Cheng, University of British Columbia (Julian.Cheng@ubc.ca)

Optical Wireless Indoor Positioning: State of the Art, Challenges, and Opportunities

Due to the increasing demands for indoor location-based applications and services such as indoor tracking and navigation, sports management, and commercial advisement, wireless indoor positioning has attracted immense research interests from both academia and industry. Recent research and development have shown that visible light positioning systems based on LED lighting is a viable technology. Such systems can achieve accuracy of centimeters with simpler infrastructure and higher energy efficiency compared with RF based indoor positioning systems. In this talk, we will provide a brief survey on the state-of-the-art optical wireless indoor positioning systems and state the practical challenges.

Xing Li, Carnegie Mellon University (xinli@cmu.edu)

Self-Healing of Analog/RF Circuits: A Statistical Approach

Sudhakar Pamarti, University of California, Los Angeles (spamarti@ee.ucla.edu)

Zero Voltage Switching Techniques for Efficiency Improvement in Backed-Off Power Amplifiers

In spite of several recent advances, the efficiency of most power amplifier architectures still degrades severely under back-off conditions. This talk describes a new approach to minimizing the efficiency degradation of class-E power amplifiers under back-off. The approach employs a combination of pulse width and load modulation under digital control to ensure that the class-E PA operates close to optimum efficiency considerations. Recent results from discrete and integrated power amplifier prototypes based on the new technique will be presented, strengths and weaknesses of this approach will be discussed and potential research directions identified.

Arnaud Werquin, Mediatek (arnaud.werquin@mediatek.com)

Multi-path Architectures for Digital RF Transmitters

Nowadays, handheld devices need to support several standards with more and more spectral constraints for spectral efficiency. This results in the use of many chips and passives. Software Defined Radio is a good candidate in order to support many standards while keeping the area small by reconfiguring the front-end and re-using RF blocks. This is mainly done by extending the digital path up to the antenna. However, the RF blocks need to be implemented in a digital way and translate the baseband filtering to RF.

A new architecture taking advantage of digital processing in multiple-path system is presented.

COFFEE BREAK (GEORGIA FOYER)

****

Takakuni Douseki, Ritsumeikan University (douse@se.ritsumei.ac.jp)

Wireless Sensor Application with Micro-Scale Energy Harvesting

Micro-scale energy harvesting that utilizes ambient energy sources makes it possible to create self-powered systems without a primary battery and normally-off sensor switches without an external mechanical ON/OFF switch. We describe two applications of a self-powered wireless sensor system: a urinary-incontinence sensor with a urine-activated battery, and a plant health-monitoring sensor with a sap-activated battery. We also describe two applications of a normally-off sensor switch that utilizes ambient light energy: a wireless mouse with an automatic switch powered by an oval-type red LED, and an IR-controlled remote-control car with a normally-off wake-up receiver powered by an oval-type IR LED.

Danijela Cabric, University of California, Los Angeles (danijela@ee.ucla.edu)

Spectrum Sensing Algorithms and Architectures for Cognitive Radios

Motivated by the spectrum scarcity problem, Cognitive Radios (CRs) have been proposed as a way to opportunistically allocate unused spectrum licensed to Primary Users (PUs). In this context, Secondary Users (SUs) sense the spectrum to detect the presence or absence of PUs, and use the unoccupied bands while maintaining a predefined probability of misdetection. Further, modulation classification, the process of identifying the modulation class employed by the transmitter, helps identify who is occupying a given frequency band and distinguish PUs from other users and interference.

Sensing wideband channels increases the chance of finding unoccupied spectrum, and therefore increases the throughput of the CR network. The narrow-band approach for wideband spectrum sensing involves sensing multiple narrow-band channels either simultaneously, or in parallel. Both solutions either require additional receiver chains, or incur additional delays in the sensing results. Therefore, sensing wideband channels in an energy-efficient manner is highly desirable.

However, wideband spectrum sensing unfolds many challenges that have not been considered in the past. In this work, we aim at answering the following fundamental question: What are the limitations of wideband spectrum sensing and modulation classification, and what are the methods to overcome them?

We propose low-complexity algorithms for both cyclostationary spectrum sensing and modulation classification. Further, as a result of wideband sampling, high-rate and high-resolution ADCs become hard to design and are power hungry. As a solution to that, we propose a method to perform cyclostationary spectrum sensing using compressive sensing measurements. Moreover, given that the sampling rate of the sensing radio cannot be adapted to all signals being sensed, we study the impact of sampling clock offsets and imperfect estimates of transmit parameters on cyclostationary features. This comprehensive system level study is vital to the robust and energy-efficient design of wideband spectrum sensing and modulation classification engines for CRs.

Harish Krishnaswamy, Columbia University (hk2532@columbia.edu)

Active Self-Interference Cancellation and Filtering for Frequency-Division-Duplexing and Co-existence in Reconfigurable RF Radios

There has been significant recent interest in flexible radios that can operate across multiple modes, bands and standards. Due to the lack of flexible electro-magnetic/electro-acoustic interfaces, such radio receivers must exhibit extremely high linearity due to the lack of front-end filtering of interference signals. While significant progress has been made in the design of ("SAW-less") software-defined radio (SDR) receivers that exhibit resilience to powerful out-of-band continuous-wave blockers, there is still considerable ground to cover in the design of reconfigurable radios that can operate in frequency-division-duplex mode or co-exist with powerful transmitters in the same platform. Such receivers must operate in the presence of powerful modulated transmitter leakage, which is presents a linearity requirement that is several orders of magnitude more challenging that the state of the art. This talk will review recent developments at Columbia University towards frequency-division duplexing and co-existence in reconfigurable RF radios. Specifically, techniques to reduce out-of-band emissions in "digital" transmitters and low-noise active self-interference cancellation techniques in broadband reconfigurable receivers will be covered.

Track R: Materials

Session R1: 2D Materials Beyond Graphene

Joshua Goldberger, Ohio State University (goldberger@chemistry.ohio-state.edu)

Group IV Graphene Analogues

The surface ligand-terminated Si, Ge, and Sn graphane analogues are a unique class of two-dimensional materials in which the electronic, optical, and thermal properties of the material can be tuned using covalent chemistry. These materials are not only promising building blocks for a variety of semiconductor applications, but also offer the potential to exhibit novel physical phenomena including 2D topological insulator. Here, we will describe our recent efforts on the synthesis, properties, integration, and device applications of these emerging materials.

Joanne Huang, Synopsys (Juan.Huang@synopsys.com)

with V. Moroz

Entrance Exam for Transistor on Two-Dimensional Materials

So far, transistors for the mainstream logic chips have evolved from planar Si MOSFETs into bulk FinFETs and are expected to keep evolving further into gate-all-around nanowires and then back into planar MOSFETs, but this time based on two-dimensional materials. FinFETs did surpass planar Si MOSFETs at 22 nm node. Nanowires are expected to beat FinFETs around 5 nm node, but it is not clear at which technology node two-dimensional materials become competitive with nanowires. In this work, we benchmark nanowires at different design rules to determine performance requirements for the two-dimensional materials to be considered a worthy contender.

Yong Chen, Purdue University (yongchen@purdue.edu)

Thermoelectric Transistors Based on 2D Materials

I will present our recent experimental studies of electrical and thermoelectric transport in gate-tunable field effect devices made from few-layer transition metal dichalcogenides (TMDC such as MoS2) and topological insulators (such as those based on Bi2Te3). These measurements help elucidate the electronic properties of these materials, demonstrate using electrical gate-tuning of chemical potential to tune (and enhance) their thermoelectric power factor, and show promises of TMDCs for thermoelectric applications.

Alessandro Molle, Consiglio Nazionale delle Ricerche (alessandro.molle@mdm.imm.cnr.it)

Synthesis and Integration of Two Dimensional Si Nanosheets in Electronic Devices

Silicon at the two dimensional (2D) limit is a target material in nanoelectronics that would naturally access a Si-based VLSI technology and its related host applications. For an ultimate scaling of silicon, the synthesis of epitaxial silicene (i.e. the graphene counterpart of silicon) on Ag-based substrates and of silicon nanosheets on MoS2 substrates is here described in the structural details. Upon delamination from the substrate, an encapsulated silicene sheet is integrated into a bottom gate field effect transistor operating at room temperature and showing ambipolar behavior. Heterosheet junction transistors based on a 2D Si monolayer on MoS2 substrates are also discussed.

Ken Shih, University of Texas (shih@physics.utexas.edu)

STM on 2D Materials

Scanning tunneling microscopy and spectroscopy (STM/S) investigation of transition metal dichalcogenides (TMDs) will be presented. More specifically, I will show that not only the quasi-particle band gaps but also the critical point energy locations and their origins in the Brillouin Zone can be revealed. Using this new method, we unravel the systematic trend of the critical point energies for TMDs due to atomic orbital couplings, spin-orbital coupling and the interlayer coupling. Moreover, by combining the micro-beam X-ray photoelectron spectroscopy and STS, we determine the band offsets in planar heterostructures formed between dissimilar TMDs (MoS_2, WSe_2, and WS_2).

Session R2: 2D Materials Beyond Graphene

Lance Li, King Abdullah University of Science & Technology (lance.li@kaust.edu.sa)

Emerging Applications of Transition Metal Dichalcogenides

Transition metal dichalcogenides (TMDs) are attractive for energy and electronic applications. Our recent success in vapor phase growth of TMD monolayer has stimulated the research in CVD growth and applications. I would discuss the synthesis and characterizations of crystalline MoS_2 and WSe_2 monolayers. These layer materials can be transferred to desired substrates, making them suitable building blocks for constructing multilayer stacks. Selected emerging applications including future electronics and hydrogen generations will be introduced.

Swastik Kar, Northeastern University (S.Kar@neu.edu)

Optical Applications in Graphene and Other Layered Materials

The meteoric rise of graphene as a 2D functional material for a range of applications has paved the way for the "discovery" of a wide variety of other 2D, atomically thin, and layered materials with a broad range of electronic and optical properties. This talk will showcase some of our recent results in photonics and optoelectronics using some of these atomically thin materials, focusing on the theme of structural and functional tunability of these materials. In one part, the talk will highlight how the low density of states of graphene can be coupled with its extremely high mobility to design tunable photodetectors with extremely high responsivity values. Next, the impact of quantum size-effects that leads to widely tune the optical properties of bismuth selenide will be described. Further, the idea of 2D alloys – materials with tunable compositions in a purely 2D lattice - will be discussed. These materials and systems are just a small representative of the new field of 2D materials that is set to revolutionize materials science in the next decade.

David Cobden, University of Washington (cobden@uw.edu)

Photocurrent Generation in 2D Materials

Devices made from graphene and other 2D materials are attracting interest for photovoltaics. In them, photocurrent can be produced even by a laser focused far from any electrical contact. This can be broadly understood as a consequence of current continuity [Song and Levitov, Phys. Rev. B 90, 075415 (2014)], but a clear-cut comparison with theory has been lacking. We find that in a magnetic field additional photocurrent is produced near the sample edges due to the photo-Nernst effect, where current is generated perpendicular to the laser-induced temperature gradient (Cao et al, preprint). This phenomenon provides an unambiguous confirmation of the Song-Levitov mechanism.

Anupama Kaul, University of Texas, El Paso (akaul@utep.edu)

2D Beyond Graphene Materials: Synthesis, Characterization and Device Applications

In this talk, I will present an overview of current progress that has been made in applying the remarkable properties of carbon-based nanomaterials such as graphene and carbon nanotubes for nanoelectronics, sensing and related applications. In addition, some recent activities in graphene-like 2D-layered crystals of inorganic materials, such as the transition metal dichalcogenides, suggest they have exciting prospects for a wide variety of applications. An overview of some recent progress in the synthesis and applications of such materials will be presented, along with highlights from a US EU Workshop held on this topic recently.

Yoshihiro Iwasa, University of Tokyo (iwasa@ap.t.u-tokyo.ac.jp)

High Density Carrier Doping in 2D Crystals

2D materials offer a superior platform for field effect transistors (FETs) and their functionalities. Here we demonstrate the electric double layer transistors (EDLT), an electrochemical version of FETs, and their properties/functionalities. Owing to the extremely large capacitance of electric double layer formed on top of semiconductor channels, EDLT is capable of high density charge accumulation, 1 – 2 orders of magnitude higher than the conventional FETs and thus new functionalities which is difficult in conventional devices are now possible to realize, functionalities stemming from superconductivity, chiral light emitting transistor, and optimization of thermoelectric properties in transition metal dichalcogenides.

COFFEE BREAK (GEORGIA FOYER)

****

Valeria Nicolosi, Trinity College, Dublin (NICOLOV@tcd.ie)

Liquid Exfoliation of Layered Materials: New Frontiers Opened by the World's Thinnest Materials

Not all crystals form atomic bonds in three dimensions. Layered crystals, for instance, are those that form strong chemical bonds in-plane but display weak out-of-plane bonding. This allows them to be exfoliated into so-called nanosheets, which can be micrometers wide but less than a nanometer thick. Such exfoliation leads to materials with extraordinary values of crystal surface area, in excess of 1000 square meters per gram. This can result in dramatically enhanced surface activity, leading to important applications in microelectronics, energy storage and harvesting, composites, etc. Another result of exfoliation is quantum confinement of electrons in two dimensions, transforming the electron band structure to yield new types of electronic and magnetic materials. Exfoliated materials also have a range of applications in composites as molecularly thin barriers or as reinforcing or conductive fillers.

Here, we review exfoliation—especially in the liquid phase—as a transformative process in material science, yielding new and exotic materials from their bulk, layered counterparts. Of all 2D materials, graphene has generated huge interest in recent years due to its unique physics properties. We have shown that high-quality monolayer graphene can be produced at significant yields by non-chemical, solution-phase exfoliation of graphite in certain organic solvents.

Until a few years ago the standard procedure used to make graphene was micromechanical cleavage, which is a very low yield production method. In order to fully exploit graphene's outstanding properties, a mass production method was necessary and the development of a method to exfoliate cheap, commercial graphite in organic solvents down to large area single graphene flakes with high yield was one major achievement. Recently this work has been extended to a wide range of two-dimensional inorganic nanomaterials. These are potentially important because they occur in >100 different types with a wide range of electronic properties, varying from metallic to semiconducting. The liquid-phase exfoliation method has now been up-scaled to produce grams of a variety of exfoliated materials per day. This talk will first discuss the galaxy of existent layered materials, with emphasis on synthesis, liquid-phase exfoliation, and characterization, to finish off with some key applications recently developed in our laboratories.

Guido Burkard, University of Konstanz (Guido.Burkard@uni-konstanz.de)

Band Structure, Spin-orbit Coupling, and Nanostructures in 2D Transition Metal Dichalcogenides

Monolayer transition-metal dichalcogenides (ML-TMDCs) are truly two-dimensional (2D) semiconductors, which hold great appeal for electronics, opto-electronics, and spintronics and have been demonstrated in FETs, logical devices, and optoelectronic structures. We investigate ML-TMDCs using k.p theory aided by density functional theory (DFT), finding trigonal warping and electron-hole asymmetries. Unlike graphene, the ML-TMDCs lack inversion symmetry leading to interesting spin-orbit effects. We use the resulting effective band Hamiltonians to describe electronic properties in nanostructures, e.g., quantum dots.

Viktor Sverdlov, Technische Universität Wien (sverdlov@iue.tuwien.ac.at)

with J. Ghosh, D. Osintsev and S. Selberherr

Electron Spin Lifetime Enhancement by Shear Strain in Thin Silicon Films

Silicon, the main material of microelectronics, is attractive for spin-driven applications due its long electron spin lifetime. The main contribution to spin relaxation due to phonons? Scattering between the non-equivalent valleys is eliminated by lifting the valley degeneracy. However, in silicon-on-insulator structures the spin lifetime is reduced due to interface scattering.

Uniaxial stress significantly boosts the spin lifetime in thin films. This is due the complete degeneracy lifting between the unprimed subbands by shear strain. This degeneracy was a long-standing problem in silicon spintronics. Lifting the degeneracy in a controllable way is paramount for future silicon spin-driven applications.

Hongyi Yu, University of Hong Kong (yu.hongyi1982@gmail.com)

Generating Spin/Valley Current in 2D Transition Metal Dichalcogenides

We will discuss two mechanisms for generating spin/valley current in 2D transition metal dichalcogenides: (1) The trion valley Hall current from the Berry curvature. The exchange interaction between the electron and hole induces an effective coupling of the valley pseudospin to its center-of-mass motion, resulting in a large Berry curvature. (2) The nonlinear valley/spin current from Fermi pocket anisotropy. It allows generating pure spin/valley flows without net charge current, either by an AC bias or by an inhomogeneous temperature distribution. The two mechanisms have distinct scaling with the field and different direction dependence on the field direction and crystalline axis.

Berry Jonker, Naval Research Laboratory (Jonker@nrl.navy.mil)

Direct Electrical Detection of Spin-momentum Locking in the Topological Insulator Bi2Se3

Topological insulators (TIs) exhibit topologically protected metallic surface states populated by massless Dirac fermions with spin-momentum locking – the carrier spin lies in-plane, locked at right angle to the carrier momentum. An unpolarized charge current should thus create a net spin polarization. Here we show direct electrical detection of this bias current induced spin polarization as a voltage measured on a ferromagnetic (FM) metal tunnel barrier surface contact. These results demonstrate simple and direct electrical access to the TI Dirac surface state spin system, provide clear evidence for the spin-momentum locking and bias current-induced spin polarization, and enable utilization of these remarkable properties for future technological applications.

Tony Low, University of Minnesota (tonyaslow@gmail.com)

Engineering Light-matter Interactions in 2D Materials and Their Metamaterials

Session R3: 3D Integration and Packaging

Zvi Or-Bach, MonolithIC 3D (Zvi@MonolithIC3D.com)

Monolithic 3D: The Most Effective Path for Future IC Scaling

It is well recognized that dimensional scaling has reached its diminishing return phase. The industry is now looking at monolithic 3D to be the future technology driver. Yet, until recently, the path to monolithic 3D has required the development of new transistor types and process flows. This paper will present the impact of emerging precision bonders and how it could enable monolithic 3D using the existing manufacturing line and existing process flows. Now the most effective path for future IC scaling is indeed monolithic 3D, which offers the lowest development and manufacturing cost for future ICs.

Paul Franzon, North Carolina State University (paulf@ncsu.edu)

Power Efficient Computing Using 3DIC Technologies

Mitsu Koyanagi, Tohoku University (koyanagi@bmi.niche.tohoku.ac.jp)

3D-LSI System Module for Future Automatic Driving Vehicle Fabricated by Heterogeneous 3D Integration Technology

We have fabricated 3D stacked CIS (CMOS Image Sensor) chip and 3D stacked multicore processor using the heterogeneous 3D Integration. These 3D chips are integrated on a Si interposer to achieve the 3D image sensor system module. We have confirmed excellent performance in the fabricated 3D stacked CIS chip. We have also proposed a highly dependable 3D multicore processor for automatic driving vehicle. This 3D multicore processor has a new self-test and self-repair function to achieve a high dependability. We have fabricated four-layer stacked 3D multicore processor and confirmed excellent performance in the fabricated 3D multicore processor.

Maciej Ogorzalek, Uniwersytet Jagiellonski Krakow (maciej.ogorzalek@uj.edu.pl)

3D System-on-Chip Layout Design Based on Shape Grammars

Computer-aided 3D ICs layout design requires effective search of discontinuous and large spaces of possible solutions. There are no deterministic algorithms able to perform the task. This paper presents a new approach to block-level 3D IC layout design. A simple shape grammar generates possible candidates for solutions. Design-specific knowledge is represented as goals and constraints that are both given in the form of predicates. The solutions have to satisfy a number of criteria in terms of geometry (layout and routing), mechanical and thermal properties, electromagnetic properties and many others. The solution space exploration is driven by an intelligent derivation controller allowing for reduction of the size of the solution space and finding an acceptable (quasi-optimal) solution. The proposed concept is illustrated with an example generated by a dedicated software package.

Gabriela Nicolescu, École Polytechnique de Montréal (gabriela.nicolescu@polymtl.ca)

Thermal Modelling and Simulation for 3D SoC

Three-dimensional integrated circuits (3D ICs) with through silicon vias (TSVs) have emerged as a very attractive technology for enhancing the performance of semiconductor devices. These architectures generate a high and rapidly changing thermal flux. Their design requires accurate transient thermal models, that can consider fast power variations and heterogeneous structures. Several thermal models for 3D ICs have been proposed, either with limited capabilities, or poor simulation performance.

This presentation introduces a novel technique based on the Finite Difference Method to efficiently and accurately compute the transient temperature in 3D ICs with TSVs. Our experiments show a 10x speedup versus state-of-the-art models, while maintaining the same level of accuracy. Additionally, we study the impact of the grid resolution on accuracy and we demonstrate the effect of large TSVs arrays on thermal dissipation.

Louis Hutin, CEA (Louis.HUTIN@cea.fr)

CoolCube Technology for 3D ICs: Challenges and Opportunities

3D ICs are emerging as a promising alternative path towards extending the Power/Performance/Area trade-off, while mitigating the issues of ever-increasing complexity and cost associated to transistor-centric engineering. The CoolCube technology in particular, which consists in processing transistors on top of each other sequentially, offers the advantages of ultra-high 3D contact density and transistor-level granularity for circuit partitioning. However, low temperature processing of the top stacked FETs is required to preserve the bottom FETs characteristics. CoolCube can also be leveraged as a comparatively simple solution versus a co-planar scheme for co-integrating separately optimized high mobility n- and pFETs (e.g. n-III-V/p-Ge).

Session R4: Materials

Emmanuel Defay, Luxembourg Insitute of Science and Technology (emmanuel.defay@list.lu)

with S. Crossley, S. Kar-Narayan, X. Moya and N.D. Mathur

Efficiency in Solid-state Cooling: Can Electrocaloric Materials Make a Difference?

Voltage-driven thermal changes known as electrocaloric (EC) effects are large in ferroelectric thin films near the Curie temperature, where entropic electrical phase transitions may be reversibly driven by electric fields DeltaE that approach the high breakdown fields generically associated with thin films. The thermal changes in a single film are small, but macroscopic assemblies of ferroelectric films in the multilayer capacitor geometry have been proposed for cheap, environmentally friendly and energy efficient cooling applications. However, candidate EC materials have hitherto only been analysed in terms of EC performance, i.e. the change in isothermal entropy DeltaS, the isothermal heat Q, and the change in adiabatic temperature DeltaT. Surprisingly, the corresponding electrical work W that is done when charging and discharging the host EC capacitors has been neglected. Therefore we introduce here electrocaloric efficiency eta to describe the ratio of reversible electrocaloric heat to reversible electrical work under isothermal conditions. This figure of merit permits a comparison of EC materials that does not depend on details of any refrigeration cycle, i.e. the type of cycle, the hot and cold temperatures of the EC material, and the sink and load temperatures. We also introduce a way to substantially improve the overall efficiency of EC coolers through the use of simple electronic components. Optimal cycles are then compared with other existing cycles, namely vapour-compression (standard fridges), magnetocaloric and elastocaloric. This analysis should in future guide the selection of electrocaloric, elastocaloric and magnetocaloric materials for novel cooling devices that are energy efficient.

Yoriko Tominaga, Hiroshima University (ytominag@hiroshima-u.ac.jp)

Crystal Structure of Low-Temperature-Grown In0.45Ga0.55As

Using X-ray diffraction and Rutherford backscattering spectrometry (RBS), we investigated the crystal structure of low-temperature-grown (LTG-) In0.45Ga0.55As on InP substrate grown by molecular beam epitaxy. Our RBS angular scan suggested that LTG-In0.45Ga0.55As grown at 220°C likely included 40% interstitial In atoms while it maintained a zinc blende structure. Our X-ray reciprocal space mapping showed that this LTG-In0.45Ga0.55As layer was compressively strained, although its lattice constant parallel to the substrate surface was smaller than that of the InP substrate. Moreover, there was no distinct transformation of this lattice distortion after annealing the sample to 550°C.

Guangrui (Maggie) Xia, University of British Columbia (gxia@mail.ubc.ca)

Mass Transport Phenomena in SiGe Systems

In the past several decades, due to the compatibility with silicon processing and the capability of mobility, strain and energy bandgap engineering, SiGe, SiGe:C and Ge have been widely studied for device applications such as metal oxide semiconductor field effect transistors (MOSFETs), tunnel FETs, hetero-junction bipolar transistors (HBTs), SiGe quantum wells and dots, modulators, and Ge lasers. When the base material changes from Si to SiGe, SiGe:C or Ge and often with the introduction of stress and dislocations, many mass transport behaviors change including dopant diffusion, segregation, activation and Si-Ge interdiffusion. The addition of carbon, dislocations, stress and Ge makes the physical picture quite complicated. This talk will try to review the research efforts and results in the above areas, which are relevant to SiGe or Ge based device structure design, epitaxial growth, doping and thermal processing of these devices.

Joerg Appenzeller, Purdue University (appenzeller@purdue.edu)

Transport in Three-Terminal Transition Metal Dichalcogenide Devices

With low-dimensional systems as 2D materials offering novel opportunities for future nanoelectronics applications, the quest is to evaluate their unique properties and potential intrinsic performance benefits. While the most commonly studied 2D system, graphene, has excited scientists because of high achievable carrier mobilities, the absence of a band gap makes many device related applications very challenging. On the other hand materials as MoS_2, $MoSe_2$, $MoTe_2$, or WSe_2, that we have carefully explored experimentally, offer sizable bandgaps at mobilities that cannot be achieved in bulk materials that are scaled down to similar body thicknesses.

In my presentation, I will first discuss the benefits of an ultra-thin body structure for scaled tunneling FET applications including tunneling devices. Source/drain contacts play a crucial role in this context and can easily mask the intrinsic performance of TMDs as will be discussed based on experimental Schottky barrier tunneling data from various TMD field-effect transistors. A careful analysis of different material systems reveals details about Schottky barrier heights for electron and hole injection as well as the band gap. These findings are then put into the context of channel length scaling and layer thickness dependence of three-terminal TMD devices based on MoS_2 transistors. Next, experimental data on the band-to-band tunneling in partially gated WSe_2 device structures will be discussed and projections about the potential usefulness of TMDs for tunneling device applications will be made. Last, a recent study on the tunability of short-channel effects in MoS_2 devices with varying channel length and body thickness will be presented.

Joshua Robinson, Penn State University (jrobinson@psu.edu)

Synthesis and Properties of Heterogeneous Atomic Layer Stacks

The isolation of graphene constituted a new paradigm in next generation electronic technologies, and even though graphene is considered transformational, it is only the "tip of the iceberg." Transition metal dichalcogenides (TMDs) and their heterostructures could have an even greater impact on next generation technologies. Molybdenum disulfide (MoS_2) is currently a leading TMD for scientific exploration, but there are a variety of other suitable, less explored, TMDs and heterostructures that exhibit very attractive bandgaps, charge carrier effective masses, and mobilities for electronic applications. Transition-metal dichalcogenides (TMDs) in the form of MeX_2 (where Me = a transition metal such as Mo, W, Ti, Nb, etc. and X = S, Se, or Te) also exhibit extreme flexibility, possession of tunable band gaps, modest electron mobilities, and wide variety of band-offsets.

Synthesizing and heterogeneously combining these atomic layered TMDs to form van der Waals (vdW) solids, where each layer may be different from the previous, is a powerful way to develop novel nanoscale materials. Furthermore, having the ability to tune the physics and chemistry with atomic-level precision is the foundation for "properties-on-demand", which can have an enormous impact on current and future technologies. This talk will elaborate on recent breakthroughs for direct growth of two-dimensional atomic heterostructures (MoS_2, WSe_2, and hBN) on a graphene template, and provide evidence that graphene can be an ideal substrate for building vdW solids.

COFFEE BREAK (GEORGIA FOYER)

****

Robert Gauvin, Université Laval, Quebec Center for Functional Materials (rgauvin@cqmfscience.com)

Microtechnologies and Mechanical Cues for the Engineering of Functional Materials for Biomedical Applications

The self-assembly approach allows for the production of living tissues in vitro, using autologous cells. The properties of these tissues can be tailored using microtechnologies and mechanical stimulations, in order to guide cell proliferation and extracellular matrix (ECM) production. The present work demonstrates that it is possible to develop anisotropic living tissues or decellularized ECM sheets that can recapitulate the physiologic architecture of planar and 3D tissues.

Nevine Rochat, CEA-LETI (nevine.rochat@cea.fr)

Porous Materials Characterization by Infrared Spectroscopy

The increase of the integrated circuits pace leaded by the decrease of transistor size, induces constant material development because of integration issues in the back-end and front-end of the line (BEOL and FEOL). For example, porous SiOCH degradation by plasma processes and deep UV resist modification by wet etching can decrease the final device performances.

The use of quasi in situ Multiple Internal Reflection FTIR permits a high sensitivity characterization of these materials in contact with various gases or liquids. Our innovative FTIR setup allowed us to investigate the material degradation during wet treatments or subsequent to plasma processes.

Huilong Zhu, Chinese Academy of Sciences (zhuhuilong@ime.ac.cn)
with M. Xu

The Adjustment and Optimization of Threshold Voltage for FinFETs with Novel Ion-implantation Methods

The adjustment and optimization of threshold voltage, Vth, for bulk FinFETs with all-last HKMG process are investigated. Novel ion-implantation methods are used to 1) modify EWF in HKMG, 2) form self-aligned punch-through-stopping layer (SPTSL) in fins and 3) create halo profile in device channel. The large amount of Vth shift to band edges caused by the implantation are obtained, which is useful for lowering manufacturing costs. The formation of SPTSL can be utilized to reduce junction capacitance and then enhance device AC performance. The creation of the halo profile with vertical implantation can be used to improve short channel effects as well as increase IC integration density.

Thierry Baron, LTM-CNRS Grenoble (thierry.baron@cea.fr)

Nanomaterials and Their Integration

Scaling down of semiconductor devices is the driving force for the development of new applications (mobile phone, memory cards, sensors...). Introduction of nanomaterials will help to improve device performances and to address new applications. We will focus our attention on the elaboration and characterization of thin film and nanowires (Si, Ge, InGaAs) by CVD and MOCVD on a Si platform. Integration in demonstrators (GAA-FET, TFET, sensors) will be shown.

Acknowledgements : F. Bassani, V. Brouzet, M. Billaud, P. Periwal, J. Moeyaert, T. Luciani, Y. Bogumilowicz, M. Martin, R. Cipro, V. Gorbenko, H. Boutry, B. Salem, T. Ernst, JB Pin, E. Sanchez, X. Bao, Y. Zhiyuan, JP Barnes, P. Serre, C. Ternon, R. Alcotte

Jeanie Lau, University of California, Riverside (jeanie.lau@ucr.edu)

Quantum Transport in Few-Layer Graphene and Phosphorene Devices

Two dimensional materials constitute an exciting platform for investigation of both fundamental phenomena and electronic applications. Here I will present our results on transport measurements on high mobility graphene and phosphorene devices. In bilayer and trilayer graphene devices with mobility as high as 400,000 cm2/V, we observe an intrinsic gapped state at the charge neutrality point. Using a "new" spectroscopy technique for measuring the Landau level gaps, we demonstrate the distinct competing states at filling factor 2 and crossing between symmetry-broken Landau levels. Our results underscore the fascinating many-body physics in these 2D membranes. Finally, I will present our recent results on fabrication of air-stable few-layer phosphorene heterostructures and observation of quantum oscillations in these devices.

Manish Chhowalla, Rutgers University (manish1@rci.rutgers.edu)

with Jean Christoph Blancon

Phase-engineered Low-resistance Contacts for Ultrathin MoS2 Transistors

Ultrathin molybdenum disulphide (MoS2) has emerged as an interesting layered semiconductor because of its finite energy bandgap and the absence of dangling bonds. However, metals deposited on the semiconducting 2H phase usually form highresistance (0.7 kohm μm–10 kohm μm) contacts, leading to Schottky-limited transport. In this study, we demonstrate that the metallic 1T phase of MoS2 can be locally induced on semiconducting 2H phase nanosheets, thus decreasing contact resistances to 200–300 ohm μm at zero gate bias. Field-effect transistors (FETs) with 1T phase electrodes fabricated and tested in air exhibit mobility values of ~50 cm2V(-1)s(-1), subthreshold swing values below 100 mV per decade, on/off ratios of >10^7, drive currents approaching ~100 μA μm(-1), and excellent current saturation. The deposition of different metals has limited influence on the FET performance, suggesting that the 1T/2H interface controls carrier injection into the channel. An increased reproducibility of the electrical characteristics is also obtained with our strategy based on phase engineering of MoS2.

Session R5: Compound Semiconductors

Takayuki Iwasaki, Tokyo Institute of Technology (iwasaki.t.aj@m.titech.ac.jp)

High Temperature and High Voltage Operation of Diamond Junction Field-Effect Transistors

Diamond is a promising semiconductor material for next generation low-loss power devices due to its superior physical properties such as a wide band-gap of 5.5 eV, a high electric field strength 10 MV/cm, and a high thermal conductivity of 20 W/cm.K. In this study, we have fabricated and operated diamond JFETs with lateral p-n junctions at a high voltage of 600 V and high temperatures up to 723 K. Also, we demonstrate the heteroepitaxial growth of a diamond film on Si substrate with a 3C-SiC buffer layer for the preparation of large-area diamond substrates.

Farid Medjdoub, University of Lille (farid.medjdoub@iemn.univ-lille1.fr)

with N. Herbecq, I. Roch-Jeune, A. Linge, B. Grimbert and M. Zegaoui

Innovative GaN Power Electronics from dc to Millimeter-wave Applications

The emerging ultrathin barrier AlN/GaN novel heterostructures are extremely promising for high frequency and high power applications. For instance, state-of-the-art PAE has been achieved up to 40 GHz owing to the control of device leakage current, material and processing quality and current collapse under high electric field. Furthermore, a fully scalable local substrate removal technique has been developed in order to drastically enhance the off-state breakdown voltage of the transistors.

In this presentation, an overview of the AlN/GaN technology development for both high voltage and millimeter-wave applications will be described.

Zhaojun Liu, SYSU-CMU Joint Institute of Engineering (liuzhaojun@mail.sysu.edu.cn)

GaN-based LED Micro-displays for Wearable Applications

LED micro-displays offer diverse applications with their superior characteristics and unique performance, particularly in high light utilization efficiency, design simplicity, long lifetime, and excellent visibility under bright day-light. In this talk, we review the design and fabrication of GaN based light emitting diode on silicon (LEDoS) micro-displays by integrating monolithic LED micro-arrays and active matrix substrates using flip-chip technology. The LEDoS micro-displays have been developed in generation with increasing display resolution and scaled pixel pitch. By integrating the red, green and blue LEDoS chips using a trichroic prism and a projection lens, a full-color 3-LEDoS projector prototype has been demonstrated.

Damien Lenoble, Luxembourg Insitute of Science and Technology (damien.lenoble@list.lu)

Materials for Invisible Electronics: Challenges and Opportunities

In early 2000s, displays, solar-cells and functional windows applications have triggered the research and development of transparent and conducting materials. Fantastic improvements in producing highly-conductive n-type transparent devices have been remarkably achieved with a focus on variants of ZnO, In2O3 and SnO2 materials. However, complementary invisible electronic perspectives are still impaired by the lack of reliable, process-able, high-quality and electronically tuneable invisible p-type semiconductor materials. Considered as one of the main challenges to be solved for the emergence of truly invisible CMOS electronic devices, we discuss novel routes of materials in this field.

Session R6: Materials

Federico Rosei, INRS (rosei@emt.inrs.ca)

Multifunctional Materials for Electronics and Photonics

Armin Knoll, IBM (ARK@zurich.ibm.com)

Nanometer Precise Patterning Using Thermal Scanning Probe Lithography

Thermal Scanning Probe Lithography (tSPL) is an AFM based patterning technique, which utilizes heated tips to locally evaporate a thermally sensitive polymer. The method allows for sub 20 nm half pitch pattern fabrication in silicon, linear speeds of cm/s, and precise 3D relief patterning. We developed a write control strategy which results in an absolute depth patterning precision of about 1nm, less than the polymer chain's linear dimension. For overlay patterning we exploit the imaging capabilities of the tool to address nanoscale devices buried beneath resist layers with nanometer scale accuracy.

Guihua Yu, University of Texas (ghyu@austin.utexas.edu)

Functional Nanostructured Polymers for Energy Storage and Environmental Technologies

This talk will present a novel class of polymeric materials we developed recently: nanostructured conductive polymer hydrogels (CPHs) that are hierarchically porous, and structurally tunable in size, shape, porosity and chemical interfaces. Given advantageous features such as intrinsic 3D nanostructured conducting framework, excellent electronic and electrochemical properties, and scalable processability, they have been demonstrated useful for a number of applications in energy, bioelectronics, and environmental technologies. Several examples on developing high-performance energy storage devices and multifunctional superhydrophobic coatings for environmental cleanup will be discussed to illustrate "structure-derived functions" of this special class of materials.

Max Ryadnov, National Physical Laboratory (max.ryadnov@npl.co.uk)

Nano-precise De Novo Protein Design

UK Protein design focuses on biophysical and mechanistic aspects of biology. It aims at providing design constraints to enable complex systems that are functional at biologically relevant length scales, and otherwise are accessible only to complex subcellular architectures. Examples include nano-vectors for gene delivery, fibrillar matrices for tissue repair and membrane-active antimicrobial agents. A key factor in all such designs is their relevance to naturally occurring macromolecular materials, be these viruses, extracellular matrices or host defence systems. The main rationale is therefore to adapt and re-purpose Nature's designs, which can lead to new solutions in a variety of applications.

Eduardo di Mauro, École Polytechnique de Montréal (eduardo.di-mauro@polymtl.ca)

with C. Santato

Exploring the Eumelanin Pigment for Supercapacitor and Memory Applications

Eumelanins are biomacromolecules responsible for plants and animals pigmentation, with interesting charge carrier transport and ion storage properties for applications in environmentally friendly electronics. In particular, eumelanin exhibits hydration-dependent conductivity, interpreted as mixed ionic-electronic conduction, and metal chelation. Here we report recent advances on (i) the use of eumelanin as electrode material for supercapacitor applications, including the role of the nature and pH of the electrolyte and eumelanin morphology and (ii) the interactions of eumelanin thin films with metal electrodes (Cu,Pd,Pt, and Fe) under electrical bias in controlled humidity conditions for potential biocompatible memories.

COFFEE BREAK (GEORGIA FOYER)

****

Sorin Melinte, Université Catholique de Louvain (sorin.melinte@uclouvain.be)

Hybrid Materials for Li-ion Energy Storage Systems

Silicon is a promising anode material in lithium batteries. An approach to roll out Li-ion battery components from silicon chips by a continuous and repeatable etch-infiltrate-peel cycle is presented. Using this process we demonstrated an operational full cell 3.4 V lithium-polymer metal@silicon nanowire battery which is mechanically flexible and scalable to large dimensions. The influence of the metal shell morphology and thickness on swelling and fracture modes of the crystalline silicon nanowires was also investigated.

Jeffrey Abbott, Harvard University (jeff.t.abbott@gmail.com)

with G. Xu and D. Ham

All-electronic Graphene DNA Arrays

Field-effect transistor biomolecular sensors based on low-dimensional nanomaterials boast sensitivity, label-free operation and chip-scale construction. Chemical vapor deposition graphene is especially well suited for multiplexed electronic DNA array applications, since its large two-dimensional morphology readily lends itself to top-down fabrication of transistor arrays. In this talk, we will present field-effect transistor arrays, made from chemical vapor deposition graphene, and demonstrate two features representing steps towards multiplexed DNA arrays: sub-picomolar sensitivity and site-specific probe DNA immobilization. The use of top-down fabricated graphene as both electrode and transistor suggests a path towards all-electrical multiplexed graphene DNA arrays compatible with existing CMOS technologies.

Ji Feng, Peking University (jfeng11@pku.edu.cn)

Exploiting Elastic Strain Gradient in Low-dimensional Semiconductors

The recent emergence of ultra-strength materials, which can sustain a significant fraction of their ideal strength, provides an unprecedented opportunity for exploring the effects of elastic strain on the quasiparticle dynamics. In this talk, I will discuss our theoretical proposal of utilizing elastic strain gradient in low-dimensional optoelectronic materials. I will discuss the possibility of creating a semiclassical charge carrier potential energy surface in low-dimensional semiconductors, by introducing an inhomogeneous elastic strain field. Numerical simulations and experimental measurements are combined to demonstrate the migration of charge carriers by strain gradient in such systems. The technological implications of these findings will be discussed.

Eduardo Albanesi, CONISET-FIUNER (eduardo.albanesi@santafe-conicet.gov.ar)

with L. Makinistian, R. Oszwaldowski, C.I. Zandalazini and A.G. Petukhov

Ab-initio Modelling of Magneto-electronic Functionality in Spintronic Materials

The injection of high currents polarized in spin represents a key to the integration of the emerging spintronics. Ab initio Density Functional Theory with the Coulomb exchange U-parameter (DFT+U) uses to be the primary theory for developing novel applications. We apply the modeling to some Heusler alloys and their interfaces, by performing an exploration of the Coulomb exchange U-parameter, reconciling the results of computational predictions with experimental measurements for Co_2FeSi, showing it is not haf-metallic, as it was largely supposed. For ferromagnetic metal/normal metal (F/N) interface of the Heusler alloy $Co_2MnAl-Au$, we predict that the Co_2 termination has better performance as a spin injector. Finally, we asses magnetic/semiconductor interfaces, and preliminary modeling of magnetic quantum dots (QD) for semiconductor lasers applications, with an effective hamiltonian justified by ab initio formulations.

Guneet Bedi, Clemson University (gbedi@g.clemson.edu)

with O. Karunwi, D. Lopez-Ferber, S.A. Tria, I. Bazin and A. Guiseppi-Elie

Impedimetric Characterization of a Mesoporous Graphitized Carbon Supported Fe-Ni Catalyst Using Microfabricated IME Devices

The development of mediator interfaces is important for the fabrication of efficient biosensors, e.g., in the measurement of mycotoxins such as Ochratoxine A (OTA). Here we use the microfabricated Interdigitated Microsensor Electrode (IME) device to perform impedimetric characterization of a polyaniline(PAn)-Mesoporous Graphitized Carbon (MGC)-Prussian Blue (PB) composite that is intended to allow covalent anchorage and enhance the peroxide sensitivity of immobilized peptide-linked oxidoreductases. This paper reports the changes in impedance of the IME device arising from the multi-step surface modification process. Impedimetric characterization of the cleaned but blank IME (control), PAn-coated device and catalyst-enhanced PAn-coated device was done. Following multi-step cleaning, anodic electropolymerization of a thin layer of P(An-co-m-AAn)/PAMPSA/pNVP/MGC was achieved (+800 mV, 250s). Impedimetric characterization of this IME device resulted in a greater magnitude of impedance at low frequencies (0-100 Hz) and at frequencies >100 Hz, the magnitude of impedance followed the same trend as the blank IME device. Following cathodic electrodeposition of five alternating layers of PB; (Iron (II,III) hexacyanoferrate(II,III), Nickel hexacyanoferrate and DI water (-400 mV; 50s, 500s, 500s, 50s), the impedimetric characterization of the IME device resulted in the magnitude of impedance dropping significantly as compared to both the blank and PAn-MGC modified device in the frequency range 0-1000 Hz. For high frequencies (>1000 Hz), the magnitude of impedance was seen to be slightly higher as compared to the blank IME device and the IME device after anodic electrodeposition.